Studies in engineering education 11

Titles in this series:

1. *Standards for Engineering Qualifications* (also published in French)
2. *Social Sciences and Humanities in Engineering Education* (also published in French and Spanish)
3. *The Continuing Education of Engineers / La formation continue des ingénieurs*
4. *Staff Development for Institutions Educating and Training Engineers and Technicians* (also published in French)
5. *The Design of Engineering Curricula* (also published in Spanish)
6. *Advances in the Continuing Education of Engineers* (also published in Arabic, French and Spanish)
7. *Engineering Technicians: Some Problems of Nomenclature and Classification* (also published in Arabic, French, Russian and Spanish)
8. *Strengthening Co-operation between Engineering Schools and Industry* (also published in French)
9. *The Environment in Engineering Education* (also published in Arabic, French, Russian and Spanish)
10. *Engineering Manpower: A Comparative Study on the Employment of Graduate Engineers in the Western World* (also published in French)
11. *Structures of Technological Education and Contributing Social Factors*

Structures of technological education and contributing social factors

Edited by Wilfred Fishwick

Unesco

Published in 1988 by the United Nations Educational,
Scientific and Cultural Organization
7 Place de Fontenoy, 75700 Paris
Printed by Bietlot Frères, Fleurus (Belgium)

ISBN 92-3-102489-2

Preface

Engineering and technical education is old enough in some countries to have a history of some two hundred years, whilst in others such an educational structure has only just been created. The way technological education has developed in those countries where it is now well-established is not only of historical interest but also provides useful lessons for the heads of engineering colleges and education administrators who are in the process of developing a suitable national structure for technological education.

An examination of the education of engineers and technicians in different countries shows a great diversity, with wide variations in the duration of studies, the degrees of specialization, the relationships between graduates at the different levels of technological education, and in the career expectations of the graduates. For example, the duration of formal studies for engineers varies from three to six years, which of course poses a question which can only be answered by an examination of the factors that determine course duration. Many of these factors will be of a social character whilst others may depend on political developments and decisions.

The evolution of systems of technological education to meet the changing needs arising from social and political aspirations depends on the provision of human and physical resources, and thus evolution may take place gradually or there may be major reforms at relatively infrequent intervals. Various ways of providing technological education have developed. In some countries technological education will take place within a larger institution such as a university or a college of further education, whilst in others engineering and technical colleges will be autonomous institutions. Some institutions operate co-operative (or sandwich) types of courses where education and training take place alternately in college and in industry.

Thus the object of this book, in the Unesco series 'Studies in Engineering Education', is to present experiences of the development of engineering education in several countries. It is hoped that these descriptions will be interesting, provide guidance or be a source of ideas for educational planners and others concerned with engineering education in developing countries. This publication is also part of the long-term activity of Unesco to countribute to the progress or reform of technological education within its Member States.

Authors in various countries, selected because of their varied education systems, were asked to prepare a study or case-history of the development and present position of their system of technological education. The study was to be a frank appraisal seen in its historical perspective, and was also expected to analyse the current situation and give views or conclusions on future trends. The ability of the system to respond to changing needs was to be given attention, and relevant economic and social factors identified.

As can be seen from the contributions chosen for publication, the education systems have very different histories and have been developed within a variety of different cultures and needs. The editor of this book, Professor Wilfred Fishwick, is responsible for the choice and presentation of the material. The various studies were in differing forms and of differing lengths and in order to keep the book within a desired size and maintain some uniformity the editor has done some pruning and re-arrangement of the original material without, it is believed, omitting any important facts or views.

The editor and the authors of the case-studies are responsible for the facts contained in this book. The opinions expressed therein are not necessarily those of Unesco, nor do they commit the Organization.

Contents

* *Note.* Each case-study has its own individual table of contents.

Introduction

Economic and social conditions in a country are intimately related and both have a history. G. M. Trevelyan, in the Introduction to his book *English Social History*, states that 'without social history economic history is barren and political history is unintelligible . . .'. This present book is concerned with the education of technologists and so with a process which is closely bound to both social and economic conditions and to political decisions in the countries reviewed here.

Technological education and training, or technological formation as these processes are often called, has not existed in its present form in any country except in fairly recent years. On the other hand, many types of technological activity have existed, in all human societies of which we have certain knowledge, for the last few thousand years and must have roots in much earlier eras. For example, glass-making has been practised in some regions for at least 4,500 years, and agriculture has a far longer history. There are now available many excellent histories of technology, some of which are listed in the bibliography at the end of this Introduction. It appears that technologies probably developed in rather haphazard ways as existing needs changed and new needs arose. Knowledge and skills were handed down from generation to generation by example and verbal instruction. The actual method undoubtedly was that of a one-to-one correspondence between master and pupil in the most formal case, and by observation of workers in other situations. Of course, this method of instructing apprentices by established craftsmen has continued into the present era and will certainly continue into the next.

Writing, using pictograms or ideograms to represent things and actions, to a certain extent can be independent of the language of the reader. The rulers of China were thus enabled to pass on instructions and laws to peoples with many dialects which were very nearly a set of distinct languages. This method of writing could in principle be used to convey technological information about crafts if there was a political or social need to do so. The advent of writing in a continuous script as practised in many Eastern and European countries was a version of a spoken language and could be understood only by someone knowing that language. Such writing is relatively easy to learn and use so that writers by no means confined themselves to laws and regulations. Writers in the old Roman Empire described many crafts, particularly agricultural crafts, but these were not textbooks for workers. They were more like letters to friends and would have had to have been duplicated by hand. As regards science in western and eastern Europe of Roman and later periods the ideas of Greek philosophers prevailed and were the subject of correspondence between men of intellect. Greek philosophy like those of India and China theorized about the physical world but was hardly concerned with the practical arts.

However, it was in the Italian states and cities from about 1300 onwards that theological and metaphysical problems ceased to be the only ones on which the educated person could demonstrate his talents. Physical problems became more and more important and advances in navigation, astronomy, mathematics and gunnery were made. Between the late fourteenth and the sixteenth centuries the mathematical knowledge of the Islamic world and India had been assimilated and a large body of literature on fortifications, hydraulics, building and architecture had appeared. As shown by B. Gille in his book *The Renaissance Engineers*, there was a body of technologists who really did deserve the name 'engineer'. The processes of making things were under scrutiny and many improvements were made. These new studies spread to Germany and then France and Britain very quickly, either by correspondence or through the printed book.

It was the perfection of a printing process in 1453 by Johannes Guntenberg of Mainz that would make the printed book or pamphlet the principal method of transmitting information to all classes of society in due course. At first, since Latin was the lingua franca of Europe in which educated men in most countries could correspond and which was used by the Church, all books were printed in Latin. Soon, books began to be printed in the various vernacular languages of Europe, and this steadily increased the reading public and so the use of books. This in turn led to a very large increase in the manufacture of paper. For example, a German, Johannes Spielmann, set up a paper-mill in England in 1588 employing 600 men. The renaissance of investigation and theorizing about natural problems and practical matters spread and with it a thirst for knowledge. Books on mining were printed in Germany as early as 1505 and were followed by others on the metallurgy of iron and brass. Mining engineers migrated around Europe and encountered new problems as mines were dug more deeply. They began to be susceptible to flooding and much attention was turned to hydraulics. Pumps of the piston type and the Archimedean screw type driven by water-wheels were constructed in several countries, though they were of limited use because of the imperfections of manufacture. The spread of knowledge was stimulating invention of new practical arts and of parallel discoveries in science.

Whilst this was going on in Europe, for reasons closely connected to the social and bureaucratic structures in those regions, the rate of progress in the improvement of practical arts had not accelerated in Arab countries nor in the ancient civilizations of India and China. Thus the civilizations of the Indian peninsula and of China and the East, whilst by no means stagnant, did not match the rate of progress in science and technology existing in Europe, and it was to be a long time indeed before they did. Many inventions were made but were either not developed or not exploited to the full. The art of building in bricks and stone was, however, of great interest in all civilizations, whether in Europe, Asia or the Americas. One reason for this was that the power and prestige of a ruler could be expressed in substantial buildings. Empirical knowledge was widespread but it was in Europe with its new outlook on the world that a true science of mechanics was developed to explain the empirical practices of architects and builders.

Education as we know it today depends greatly on the use of the printed word, but during the fifteenth, sixteenth and seventeenth centuries there was no national or state system of education in any country in Europe or indeed elsewhere. In many countries religious organizations had schools which were associated with the church, mosque, or temple, whilst the wealthy provided tutors for their offspring. There was certainly no system of technological education anywhere until the seventeenth century, when some European countries set up schools for the military engineer, to be followed by more such schools in the eighteenth century. The word 'engineer' was usually restricted in its use and described a military man who built and demolished fortifications, built roads and bridges, and used explosives, battering rams and the like. The skills needed by these engineers was known and documented, but the first generals to require their engineer officers to have a special training seem to have been from Holland at the beginning of the seventeenth century. The French military *corps du génie* was formed in 1676 but there were no military engineering schools in France exclusively for military engineers until 1749 at Mezières. Teachers at artillery schools and engineering schools produced many textbooks which were available to non-military engineers. Engineering schools were founded in many countries in Europe for military purposes about this time. Even in England where, because of its isolation, the study of fortifications was of less interest than in Europe the Royal Military Engineering College was founded in 1741. In Britain there was more emphasis on engineering for civilian purposes, and indeed the term 'civil engineer' was first used there.

Long before 1700 scientific and technological advances had been taking place in many European countries and the new knowledge was disseminated not only through books but by lectures given by the intellectual élite and by lecturers who charged their audience for the privilege of listening. Many of these were itinerant lecturers moving from town to town, whilst others who were located in the larger towns often lectured to scientific clubs and societies. The audiences at these lectures did not all come from the educated middle classes but contained many artisans. The drive for self-education was very strong in Britain following the reformation of the established Church and the loosening of the bonds of the former feudal society. After the Civil War the feudal system had finally disappeared, people were free to find employment where they could, and ideas flowed freely. No longer did the monarchy, after its restoration, ever regain former powers in Britain. As will be seen in the case-history of technological education, the drive for self-improvement did not slacken and by the year 1700 Britain was in a better position to exploit new ideas than any other part of Europe and indeed of the world. There was just as much expertise and knowledge in other parts of Europe but it was in Britain that the first industrial revolution began and flourished. It was in Britain that iron became a cheap commodity and that water-power and steam-power were used widely to drive all kinds of new and old types of machines for the production of materials and goods. The social structure in Britain was able to accept and profit from these new developments.

The Industrial Revolution and education

The revolution in the production of goods and in the accompanying trade was not confined to Britain and her North American colonies but spread into other countries. However, it was not until the late nineteenth century that Germany and the United States of America finally caught up with industrial output in Britain, to be followed by others in due course. It should be noted that in the year 1700 the population of France was three times that of Britain and probably had more scientists and intellectuals, but the social system did not produce the large number of entrepreneurs that Britain did, nor had France so many literate and mobile artisans and workers. Such people are needed in industry and the more sophisticated is industry the larger the number of well-educated people required to keep it growing.

As noted above, modern industry needs literate workers and even the new industries of Britain and Europe in the earliest years of the first Industrial Revolution needed some educated leaders and some literate workers. There was no country in the year 1700 which provided education for the masses or any systematic education in science and technology, except possibly for medical practitioners. Classical education was available in several European countries for those of the middle classes who could pay fees in schools set up by the churches or by bequests from wealthy people. In many of these schools there were free places for a few pupils mostly drawn from the class of artisans and clerks. Mathematics was taught in most of these schools but little, if any, science. Mathematics was also taught in many of the numerous universities in Europe.

The first country associated with the new industry having schools supported by public funds and open to all was Scotland. Following an earlier but none-too-successful start in 1633, a large number of parish schools were opened in 1696 and the following years. These schools were associated with the Presbyterian church in the parish and supported by taxes and fees paid by all except the very poor. The towns also had supported burgh schools used mainly by the children of artisans, merchants and the town gentry. Many children from parish schools actually went on to one of the four universities in Scotland.

In 1763 Frederick William I of Prussia introduced compulsory elementary education for all children and other Germanic states followed this example. This was to mean that in about fifty years northern Germany and to a lesser extent catholic southern Germany would have the most literate population in Europe and indeed the world. This was to have a profound effect on the development of industry in Germany.

Elementary education for the majority of the population, and supported by public funds, came much later in France, Italy, the USA, Russia, and in England and Ireland. Indeed, at first sight it appears strange that Britain kept its industrial supremacy for so long without public support for the schools and colleges in England. First, there was great opposition shown by influential people to government interference in almost any activity except defence. Secondly, this independent spirit had resulted in many forms of private education including charity schools for the poorest classes which were supported, though not well, by gifts and benefactions. Also education could be obtained through comrades, friends, parents and relations. It was well noted by F. Smith in his book *English Elementary Education* that 'education is never synonymous with schooling . . .', but even so this haphazard system of elementary education failed to satisfy both the needs of the working classes and the industrial employers. A true national system was not fully introduced in England until 1870, although a form of public support for schools had been introduced in 1834.

Technological education

Probably the first institution set up specifically to educate technologists was the School of Mathematical and Navigational Sciences in Moscow. This followed the visit by the Tsar, Peter the Great, to Holland and England where he had actually worked as a shipwright. The head of this new college was Professor Andrew Farquharson of Aberdeen University who was assisted by two other Scottish professors. The initial enrolment in 1701 was 200 students, most of whom had experienced little schooling. The school was moved to the new capital at St Petersburg (now Leningrad) in 1715 and became the Naval Academy. The school in

Moscow became a preparatory school for the academy. There were practically no schools in Russia at this time, so when the Academy of Sciences was created in 1724 there was a lack of qualified candidates, and professors such as Euler, the Bernoulli brothers from Switzerland and others did nothing but research for many years. As literacy spread a School of Mines opened in 1773 and an Institute of Engineers for Ways of Communication in 1809. These institutions were primarily for military purposes and the new industrial practices were slow in coming to Russia. However, trade schools and technological institutes followed, then the Polytechnics in Kiev (1898), Warsaw (1898), St Petersburg (1902), and Novotcherkusk (1906). The industries which had finally come to Russia were set up mostly by foreign entrepreneurs or foreign firms. Since the revolution in 1918 technical and engineering education and training in Russia and the Union of Soviet Socialist Republics have expanded enormously to meet the demands of a highly industrialized country.

The skills of British engineers and workforce and the productive capacity of industry did not go unnoticed by other nations, but having different social structures or traditions they decided to train their engineers in special institutions rather than in workshops and through discussions in societies. Quite early under Louis XIV the French minister Colbert had organized state-owned factories, for luxury goods mostly, and had also caused several large canals to be dug. These and other prestigious public works led to the foundation of the Académie Royale d'Architecture in 1671 which, with its fifty students, was the first technical institute in France educating surveyors and civil engineers for the state service. A Corps des Ponts et Chaussées was created in 1716 which became the well-known École des Ponts et Chaussées in 1775. After the French Revolution the need for new educational structures was recognized and finally the École Polytechnique for the study of the mechanical arts and science was founded in 1794. It had a wonderful collection of scientists and mathematicians as teachers, many of whom came from the closed University of Paris. From then on, with some hitches, the number of higher institutions educating engineers and technologists in France steadily increased, and there are now about 150 such places, some being monotechnics and others polytechnics. Technical institutes for technicians and trade schools of various forms have kept pace with the engineering colleges and their graduates are of high quality and in large numbers.

In the German states there were many universities by 1800 and mathematics and science were recognized studies. The science of chemistry was taken very seriously and was to become of great importance to German industry in later years. There were several technical institutes but it was not until 1851 that one of them in Dresden was upgraded to become a technical university or Technische Hochschule. By 1879 eight more institutes had been upgraded to Technischen Hochschulen. These institutions were to produce a formidable galaxy of engineers and industrialists who created industries to match those in any country, and in many ways to surpass them. As the polytechnics developed, so did an excellent infrastructure for the education of technicians and craftsmen and craftswomen; this was at least as good, if not better, than that found in any other country. Germany in 1938 was the result of an amalgamation of 39 states; today there are two Germanies but the engineering traditions of the older Germany have persisted in both.

Finally, it is interesting to take a brief glance at what has happened in the United States of America which, in about 130 years, developed from being a British colony into the most powerful industrial nation in the world. Today, after a further sixty years or so, it is still in that position, though being caught up by other countries. The initial traditions and practices in education were British in origin and there was a substantial fraction of the population who believed in the moral value of education, as had their forefathers in England and Scotland. Thus several university colleges were founded even before independence and the trend continued thereafter. There were no public funds for education until well into the nineteenth century, but technical education existing in the form of mechanics institutes in Britain were replicated in America. These institutes were really societies of artisans who came together to obtain further education in science and the practical arts through discussion and by hiring lecturers. In the USA these self-governing societies were often called lyceums. The universities did not neglect science, being influenced by German and Scottish universities to some extent and by the practical outlook of the population. However, the first true engineering college was not in a university but set up by S. Rensselaer in Troy, in the state of New York, in 1823. The aim was to provide a practical scientific education for young men who had just graduated from the universities and who wished to have a career in industry. These ideas were akin to those of the great Francis Bacon of England but now almost forgotten there, having been superseded in the universities by stultifying ideas so well put forward by J. H. Newman (later Cardinal). The Rensselaer Polytechnic Institute, as it was renamed in 1849, graduated its first four civil engineers in 1835, and from then on specialized more and more in teaching engineering not only to graduates but also to direct entrants from schools. It was the forerunner of very many such private and public institutions in the USA. The public provision of higher education in the liberal arts, science and engineering was hastened by the creation of land-grant colleges in 1861 and 1862. A grant of land and some other resources had been made to each state in order to support at least one college which, amongst other things, would teach 'such branches of learning as are related to agriculture and the mechanic arts'. These land-grant colleges eventually became now-famous universities. Sometimes the grant was used to add suitable departments to existing private and state-supported colleges. The state governments were now thoroughly involved in higher education, and this was quickly followed by the giving of support to all kinds of education within their states. Primary, secondary, technical and university education flourished in a way not matched elsewhere for many years. Opportunities for technological education along with a social climate fostering innovation and risk-taking plus the inflow of capital and ambitious immigrants from Europe led inexorably to industrial supremacy, which only now is possibly under threat of being overtaken.

Recent developments and some conclusions

From the case-histories presented later and from other sources it is clear that industrialization of a country can be achieved in a variety of ways, but that no country has been able to develop modern manufacturing or service industries without external inputs of some kind. This applies just as much to the countries that first developed modern types of industry as to those that are now developing their industries. Furthermore, it is possible for industries to be introduced into a country using not only imported knowledge but also imported, or at least non-indigenous, engineers and other professionals. However, this is only a temporary phase for extensive industrialization requires a sufficient number of educated and literate people and a social climate fostering various kinds of vocational education, as for example was the case in the Republic of Korea or Taiwan.

Then, quite inevitably and necessarily, every country which wishes or needs to industrialize decides to train its own technologists and other ancillary personnel. This decision has usually been preceded by a period during which citizens go abroad in order to be educated and trained as technologists. Indeed, citizens often go, or are sent, abroad for technological education for several years after suitable educational establishments have been set up in the home country, simply because it takes a long time for new institutions to be staffed and equipped adequately and to develop a good reputation.

A natural input when starting up the higher education of technologists is a foreign model on which to base the programmes of new technological institutions such as engineering colleges. Countries which were colonies or dependencies of an occupying power, tend to get their models from those states, whilst some with a close political relationship to a larger industrialized state may draw their models from that country. If they are wise they will modify the imported system to meet their own needs. An example of such a modification is to be found in the chapter on Pakistan. The original engineering schools were faculties of the four universities following the quite common British practice. As the years went by their development, and effect on industrial development, was disappointing. Hence a bold decision was made to upgrade the four faculties into autonomous technical universities, which from that point on have developed and grown in size, with gratifying results for the industry and the general progress of Pakistan. There is a lesson here for other industrializing countries, namely that imitation flatters but does not always produce a desired result.

The historical surveys in this book give vivid pictures of the way technological education has developed in certain countries but only touch lightly on what might happen in the future. Therefore this Introduction will end with some comments on some other aspects of technological education that are likely to be of increasing importance in all countries in the years to come.

CO-OPERATIVE EDUCATION AND TRAINING

Co-operative or 'sandwich'-type education is a form of education which includes real industrial practice as an integral part of the process of obtaining an engineering degree or other technological qualification. Some periods in industry are interleaved with periods in a college or university. In a 'thin sandwich' course six months in industry are followed by six months in college, whilst in a 'thick sandwich' twelve months in an industry are followed by a year, or even two years, in college. As will be seen later in this book it is practised very successfully in Canada and in Scotland (and in the rest of the United Kingdom). It is also practised in countries such as Australia, the Republic of Korea, the USA, the Philippines, and in Singapore, to name but a few.

True co-operative education and training (CET) is an academic programme, organized and supervised by an educational institution, in which is incorporated productive industrial activities. There is thus an arranged partnership between the college, a private or public enterprise, and the student, all of whom should benefit and all of whom have responsibilities. CET as defined here does not accept periods in industry whether prior to entering college, during vacations or after graduation, as being sufficient for the education to be designated as co-operative education. In true CET worthwhile work experience is assessed and forms an important part of the final assessment which leads to some professional qualification.

Wherever it is practised, CET is recognized by industry as a form of higher education that produces engineers and other technologists of high quality, who can undertake productive work without too much readjustment. If this is so it might be asked why it is that the formation of all engineers and other technologists does not conform to this model. The answers to this question vary from country to country, but some general observations can be made.

First, it is necessary to realize that CET demands a close partnership between engineering colleges and industry, based on a realization that the colleges exist to satisfy some of the needs of industry, and that these needs will not be met unless industry is prepared to involve itself with educational institutions. In particular, industries have to accept students for periods of from six months to one year under rather stringent conditions. These students form two groups: one group having been initially recruited and employed by industry before going to college; and the second group consisting of students who have been admitted to college without any prior industrial affiliation.

It is relatively easy to return students in the first group to their sponsoring industries for supervised and assessed experience—although it might be better for the student to work elsewhere—but in many countries and especially in developing countries it is not always easy to place all members of the second college-based group. It is a demanding task for the academic staff to arrange and supervise students in sponsoring organizations and this task becomes even more onerous if they have to find places in industry for non-sponsored students. It can be and is done by many colleges, but nevertheless it is a factor preventing the adoption of CET in many other colleges.

Second, many engineering colleges are in fact faculties or departments in multi-disciplinary universities where the majority of students do not have structured programmes incorporating academic and non-academic work. The introduction of CET for engineers is resisted because the course would be lengthened and this might, it is feared, reduce the quality of students opting to take engineering. This belief is a very important factor in countries having one or two universities which are considered superior to others and whose engineering schools do not practise CET. If they did practise CET then it is certain all other engineering schools would follow their pattern. However, complacency even in the 'best' universities is not easily overcome.

It might be thought that in countries where all or nearly all engineers and technologists are educated in autonomous technical universities or colleges that true CET with all its advantages would be universal, but this is not so. It appears there are no countries where the majority of technologists undergo CET, and the majority of engineering schools practise CET. Once a form of education has a well-established pattern it is quite difficult for the system to change by internal action. External action is usually needed.

CONTINUING EDUCATION AND DISTANCE-LEARNING

The case-histories presented later discuss the evolution of traditional institutions offering courses leading to first, and perhaps more advanced, qualifications following full-time studies. However, it is realized that today almost all technologies are changing, so that any practising technologist must keep himself up-to-date. When technologies changed slowly this could be accomplished by reading a few journals, and a book from time to time. Today most engineers and applied scientists have to acquire new theoretical and practical skills in order to solve their problems arising in their current work, or to move into a new field of technology, and this process must be done both quickly and efficiently. Industrial progress depends on innovation which in turn often depends on the use of new knowledge and techniques. Thus the continuing education of engineers (CEE) and others is now a very important activity in several countries, and indeed is spreading rapidly to all countries. Schools of engineering and applied sciences offer courses to practising technologists, sometimes in rather general topics and sometimes in very specialized subjects. In some countries there are also private organizations that offer similar courses. Commonly an on-campus course will last five days and participants, or more likely their employers, will pay a fee. These courses are often residential and can attract technologists from a wide area.

There are also in-company courses given to larger organizations and enterprises which require some form of training on a more private basis, and have a group of technologists needing more or less the same training.

Many universities and colleges in the industrially developed countries have departments of continuing education whose function is to organize on-campus and off-campus courses utilizing tutors drawn from the college or from elsewhere. In some institutions this is now quite a big business. Several private or semi-private organizations in the USA, and Europe, buy course material from wherever it can be obtained and will provide courses tailored to suit any customer who can pay. Many governments provide CEE, especially in the new areas of high technology such as computing, information technology, computer-aided design and manufacture and bio-engineering.

Not all engineers can or want to attend courses away from their place of work or home. There is thus a growth in so-called distance-learning (DL) whereby the participants can study and learn at work, or at home, or at least near home. There are a large variety of forms that the instruction can take, many of which are based on the systems set up by the so-called Open Universities. Indeed these Open Universities, of which there are now more than one hundred scattered around the world, were originally founded to enable people of any age to obtain a first degree or similar qualification without attending a college or university, but they are now turning their attention to the CEE market.

Distance-learning is carried out using correspondence, books, local tutors, local community colleges, possibly radio and/or television, audio tapes and video tapes, and recently via the telephone using modems connected to home computers. For long courses summer schools or their equivalents are also found very useful. Now, satellites are being used for dissemination of instruction over very wide areas. At first this was confined to North America but now satellites are being used experimentally to bring courses on technology to Europe. These developments in the techniques of distance-learning will undoubtedly affect all types of higher education and especially vocational education in the future. It is not difficult to imagine that the distinction between on-campus education and off-campus education will become blurred. Perhaps most types of technological education and training will become co-operative and modular in form, with a final qualification being awarded when an assigned number of modules, academic and non-academic, has been accumulated. Certainly there are changes ahead and therefore much scope for innovation, and maybe these innovations will arise in newly developing countries.

Nevertheless, I am sure that there will be no one superior system of technological education which will be discovered and used by all. Differing societies with differing objectives will need different forms of education and training, as seems to be the case today.

Bibliography

ARMYTAGE, W. H. G. *The Rise of the Technocrats*. London, 1961.
——. *A Social History of Engineering*. 3rd ed. London, 1970.
ARTZ, F. B. *The Development of Technical Education in France, 1500 to 1850*. Cambridge, Mass., 1966.
DERRY, T. K.; WILLIAMS, T. I. *A Short History of Technology*. Oxford, 1960.
EMERSON, G. S. *Engineering Education, A Social History*. 1973.
FLEMING, A. P. M.; BROCKLEHURST, H. J. *A History of Engineering*. London, 1925.
GILLE, B. *The Renaissance Engineers*. 1966. (Translated from French.)
HERFORD, C. H. *Germany in the Nineteenth Century*. Manchester, 1915.
MCGIVERN, J. G. *One Hundred Years of Engineering Education in the United States, 1807–1907*. New York.
MUSSON, A. E.; ROBINSON, E. *Science and Technology in the Industrial Revolution*. Manchester, 1969.
NEEDHAM, J. *Science and Civilisation in China*. Cambridge, 1954.
SINGER, C.; HOLMYARD, E. J.; HALL, A. R. *A History of Technology*. Oxford, 1954–58.
TIMOSHENKO, S. P. *Engineering Education in Russia*. New York, 1959.
VEBLEN, T. *Imperial Germany and the Industrial Revolution*. New York, 1915.

Case-study for Canada

Glenn A. Morris,
Associate Dean, Faculty of Engineering,
University of Manitoba, Canada

Contents

The education and practice of engineers and technologists in Canada

Introduction

Canada, with a population of 25 million people and a land mass of 9.9 million square kilometres, is a federation of ten provinces, each with its own provincial legislature, and two territories. As illustrated in Figure 1, the provinces vary greatly in both size and population. While large and resource-rich, Yukon Territory and the Northwest Territories are very sparsely populated. Indeed, over 90 per cent of Canada's population is located within 200 miles of the Canada–United States border.

The Dominion of Canada was formed through the confederation of the provinces of Nova Scotia, New Brunswick, Ontario and Quebec, by the British North American Act, passed by the British Parliament in 1867. Subsequently, the four founding provinces were joined by Manitoba, British Columbia, Prince Edward Island, Alberta, Saskatchewan and Newfoundland. In 1982, the Canadian Constitution was patriated and it is now the responsibility of the Federal Government in Ottawa.

Under the Constitution, responsibility for legislation in the fields of education and the governance of the various professions is assigned to the provinces. That fact has had a profound effect on the evolution of the structures and organizations that have been developed for the education of, and the regulation of the practice of, engineers and technologists.

As might be expected, the provinces have jealously guarded their rights in the field of education. On the other hand, the Federal Government has attempted to pursue national policies in areas such as immigration and manpower training, industrial strategy and regional economic development: policies that are heavily dependent upon the provincial educational systems. Consequently the Federal Government, through a variety of cost-sharing programmes, has tended to pursue its strategies while simultaneously promoting consistency among the various provincial education systems. A consequence of this dispersion of responsibility has been the absence of an articulated national policy on higher education. Notwithstanding that fact, there is a remarkably high degree of uniformity throughout Canada in the education of engineers

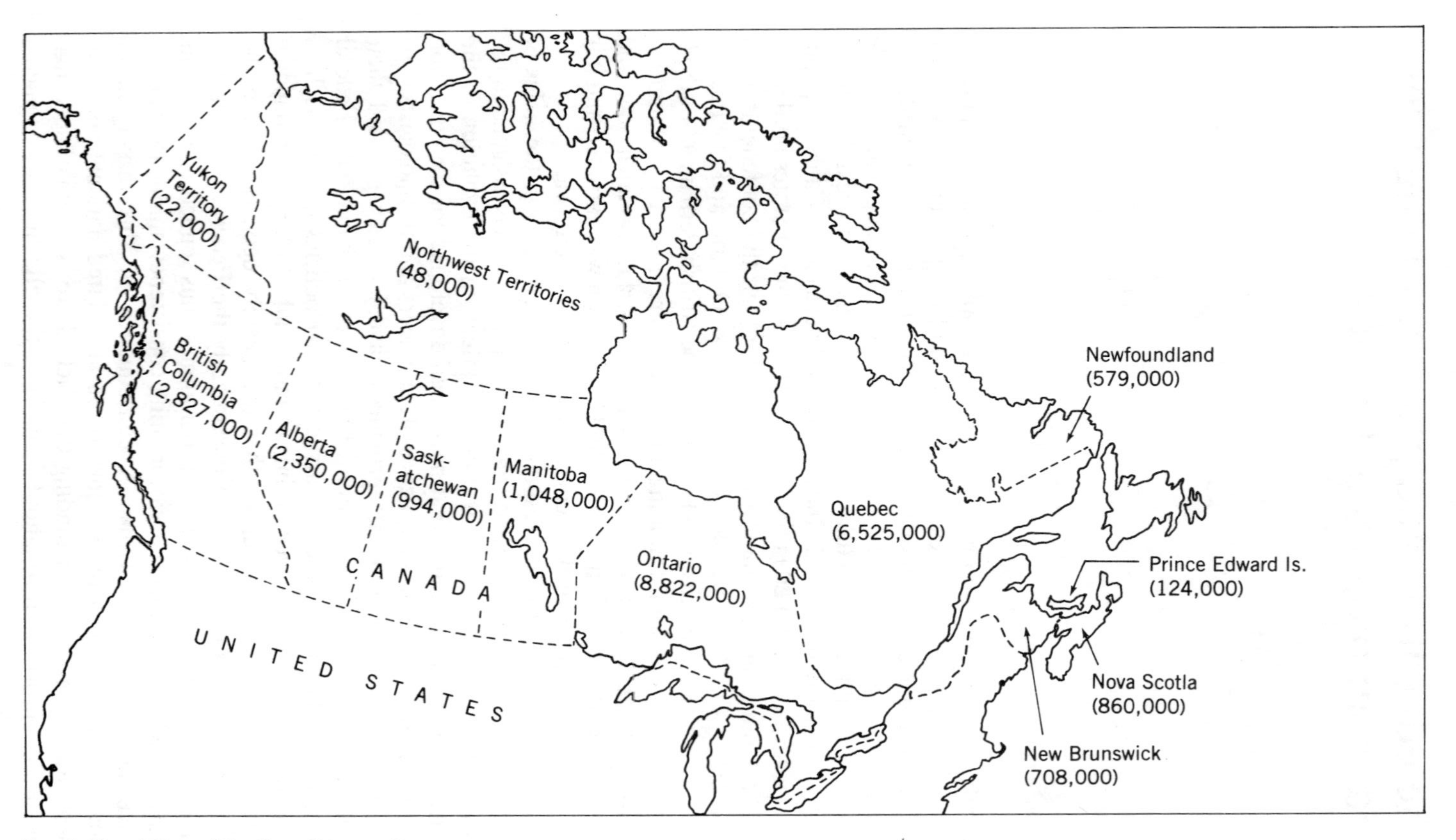

FIG. 1. Population of the Canadian provinces.

and technologists. Curriculum requirements, quality and level of instruction, duration of programmes and quality of physical plant are more consistent than in most countries.

Likewise, the conditions under which engineers and technologists practice are highly consistent. While an autonomous Association of Professional Engineers in each province and territory oversees and polices professional practice, the associations belong to a federation that encourages uniformity of standards and practice. Following the lead of the engineers, engineering technicians and technologists have also formed provincial societies of certified engineering technicians and technologists which in turn created a federal body.

The result is a consistently high quality of education of engineers and engineering technologists and an assurance of a good standard of practice. Almost all of the practising engineers and a large proportion of the practising technologists and technicians are members of the organizations that regulate and police practice in the engineering fields in Canada.

The Canadian Constitution recognizes two official languages: English and French. Approximately 5 million Canadians, most of them living in Quebec, use French as a first language. Nevertheless, the majority of Canadians function in only one of the two official languages. Many of the engineering and technology educational institutions in Quebec use French as the language of instruction and almost all of the rest in Canada use English.

Educational levels

Education in Canada can be classified as elementary, secondary, post-secondary, tertiary and postgraduate. Elementary education normally begins at the age of 6 and comprises grades 1 to 6. In most provinces, secondary education comprises junior high school (grades 7, 8 and 9) and senior high school (grades 10, 11 and 12). A very large majority of elementary and secondary schools are publicly funded and administered by local school boards, although there are private non-sectarian and Church-sponsored schools. At the elementary level education is general and basic. At the secondary level the student can choose between an academic or a vocational programme. While a few high schools in metropolitan areas are oriented mainly toward vocational training, most are 'composite' schools which offer both academic courses preparatory to university and vocational courses which prepare a student for an occupation or for non-university post-secondary education.

In Canada, the term 'university' is used to describe any post-secondary institution that has been given the power to grant degrees. With two exceptions, engineering programmes in Canada leading to the baccalaureate degree are offered by university faculties. The exceptions are the

École Polytechnique, an engineering educational institution that is affiliated with the Université de Montréal, and the Royal Military College, a federal educational institution for military personnel. Most of the institutions that offer engineering baccalaureate programmes also offer postgraduate programmes at the master's and doctorate levels.

Programmes leading to the diploma in engineering technology are normally offered by non-degree-granting community colleges. While the technology programmes are post-secondary, high school graduation being a normal entrance requirement, the colleges normally also offer tertiary level vocational programmes. The latter do not have high school graduation as a prerequisite. However, they are normally quite specialized, incorporating material not included in the high school vocational programmes.

The education-practice system

The system of education for engineers and technologists in Canada is influenced to some extent by the organizations that govern the practice of engineering and technology. Furthermore, the formal education system is complemented by a group of engineering technical societies and a national organization that is the primary source of funding for engineering research in the educational institutions. Hence, the description of the educational system includes a consideration of the following elements.

The faculties of engineering
With two or three exceptions formal engineering education is accomplished in university faculties or schools of engineering or applied science. While the requirement for the practice of engineering is the Bachelor's degree, most engineering faculties also offer programmes leading to the Master's and Doctor of Philosophy degrees.

Engineering research
In 1916, the Federal Government established an Honorary Advisory Council to foster and co-ordinate research activity. From its inception, the council maintained a programme of university research grants and postgraduate scholarships. Following a name change to the National Research Council in 1925, the council developed an in-house research capability to complement its support of university research. In 1978 the programmes of university research grants and support for the training of research manpower were assigned to a newly created agency, the Natural Sciences and Engineering Research Council.

As a primary source of financial support for engineering graduate students and engineering research, the council has a strong influence on the environment in Canada's engineering faculties.

The technology educational institutions
Prior to the 1950s, technical education in Canada tended to diverge into two streams, either university level education or vocation/trade school education. Following the Second World War there was an influx of engineering technical personnel from abroad with academic technology backgrounds. Soon a need was perceived for institutions to provide engineering technological education at a level between those provided by universities and the vocational schools. Beginning in the early 1960s, such institutions were established in all provinces.

The engineering technical societies
The system of formal education of engineers is complemented by a group of engineering technical societies whose function is to provide forums for the development and exchange of engineering knowledge and to address the continuing education needs of the engineering community. The oldest Canadian technical society, the Engineering Institute of Canada, was established in 1887. In 1970 it became an umbrella organization for a group of specialized constituent technical societies.

In addition, British, American and international technical societies have substantial numbers of members in Canada. Examples are the Institution of Civil Engineers, the American Society of Mechanical Engineers and the Institution of Electrical and Electronics Engineers. Generally, there is close co-operation and often a reciprocal agreement between a Canadian society and a corresponding international one.

The provincial Association of Professional Engineers
Each province and territory in Canada has its Association of Professional Engineers. The practice of engineering in any province is governed by, and restricted to, members of the provincial association. The requirements for membership include suitable academic preparation. While there is no formal connection between the various faculties of engineering and the associations of professional engineers, the design of engineering curricula is influenced by the academic requirements of the associations.

The Canadian Council of Professional Engineers and the Canadian Accreditation Board
The Canadian Council of Professional Engineers is a federation of the associations of professional engineers in Canada's ten provinces and two territories. The council deals with matters of mutual concern to its constituent members. One of its main functions, the accreditation of engineering programmes, has been delegated to one of its standing committees, the Canadian Accreditation Board. The board conducts a programme of periodic accreditation visits to Canadian engineering faculties and it publishes annually a list of accredited engineering programmes. Graduates of those programmes are considered by the

provincial associations of professional engineers to satisfy the academic requirements for membership.

The societies of certified technicians and technologists
During the 1960s groups of engineering technicians and technologists in the various Canadian provinces organized and incorporated societies of engineering technicians and technologists to regulate the practice of technology. Then in 1972, following the example of the Canadian Council of Professional Engineers, the provincial societies formed their own national organization, the Canadian Council of Engineering Technicians and Technologists. That body created the Canadian Technology Accreditation Board which currently accredits engineering technology programmes in some of the provinces.

The beginnings of engineering in Canada

Engineering work in Canada dates back to the seventeenth and early eighteenth centuries, when French military engineers constructed fortifications to defend their settlements against the incursions of the Indians and the British colonists to the south. Later, the French were succeeded by the British military engineers, who not only erected fortifications but also built roads and undertook the construction of canals and other public works. The first road was opened from Quebec to Montreal in 1734 and the St Lawrence canal system was started in 1779. A highway from York to Simcoe was constructed in 1794. The canal system, which included the present Lachine, Welland and Rideau canals was under active development during the beginning of the nineteenth century. Many of the early roads and canals were designed for military purposes but they subsequently proved invaluable in the development of communications in Upper Canada (the present Ontario).

Railway construction started in Canada in about 1830 and the first line, that from Laprairie to St Johns, Quebec, was opened in 1836. As railway construction increased, there was an influx of civil engineers, primarily from Britain and the United States. The railway from Montreal to Toronto was completed in 1856 and when the Victoria Bridge in Montreal was built, in 1860, there was a 1,350-km railway link stretching from Portland, Maine to Sarnia, Ontario.

In 1867, the Confederation agreement created the Dominion of Canada as a federation of the provinces of Ontario, Quebec, Nova Scotia and New Brunswick. It also called for a transcontinental railway to link the eastern provinces with the Pacific. The construction of that railway is perhaps the most spectacular engineering enterprise ever undertaken in Canada. While contending with immense financial problems, an armed rebellion, bitter political controversy and the rigors of the Selkirk and

Rocky Mountain ranges, the builders flung about 3,300 km of steel across a continent in a mere five years—exactly half the time stipulated in the contract. In the period between 1881 and 1885, Canada was forged into one nation by the building of the Canadian Pacific Railway.

Engineering education

As illustrated in Table 1, the history of engineering education in Canada is quite short. At the beginning of the twentieth century, only five Canadian universities offered engineering instruction in six speciality areas.

Moreover, except for the engineering faculties established in the new provincial universities in the four western provinces in the first decade of this century, no new engineering educational institutions were developed until just prior to the Second World War. This slow development was compensated for by astonishing growth during the 1950s and 1960s. During a twenty-year period beginning in 1948, no fewer than eighteen Canadian universities established faculties to teach engineering. Today, twenty-nine universities offer engineering education under more than forty programme titles.

THE EARLY YEARS

The University of Toronto

The initial step to provide formal engineering education in Canada was taken at the University of Toronto in 1851. The importance of science and technology in the development of the country's resources had become widely recognized. While there was an obvious need for well-trained engineers to work on the construction of canals, harbours, roads, railways and bridges, there were fewer than a hundred engineers in the entire country.

Under the Baldwin Act of 1849, Toronto's King's College, which had been established in 1843 by the Church of England, was transformed into the non-sectarian University of Toronto. In January 1850, a commission was empowered to establish new areas of study and by mid-1851 it had invited applications to fill a Chair of Civil Engineering. Seven applications were received, but no selection was ever made. The university proceeded with plans for engineering instruction, but it was not until 1859 that the first four students were admitted.

The curriculum included mathematics and natural philosophy; English; French; history; chemistry and chemical physics; mineralogy, geology and physical geography; geodesy and drawing; and civil engineering, including the principles of architecture and engineering finance,

TABLE 1. Years of establishment of engineering institutions and programmes

	Institution	Engineering Unit	Agricultural/ Bio Resources	Chemical	Civil	Computer	Electrical	Physics	Geological	Industrial	Mechanical
Alberta	1906	1913		1926	1908	1980	1915				1959
British Columbia	1908	1915	1947	1915	1915		1924	1943	1921		1940
Calgary	1966	1967		1967	1967		1967				1967
Carleton	1942	1957			1957	1982	1957				1957
Chicoutimi	1969	1969							1969		
Concordia	1963	1963			1966	1981	1966				1966
École Polytechnique	1873	1873		1958	1873		1958	1958	1958	1971	1958
Guelph	1964	1964	1964								
Lakehead	1965	1965		1972	1972		1972				1972
Laurentian	1960	1960		1960	1960						1960
Laval	1852	1937		1941	1946		1942	1957	1945		1948
McGill	1821	1878	1971	1908	1872		1891				1883
McMaster	1887	1957		1957	1959	1971	1957	1957			1957
Manitoba	1877	1907	1968		1907		1907		1950		1947
Memorial	1949	1949	1949		1969		1969				1969
Moncton	1963	1957			1968					1972	
New Brunswick	1859	1854		1961	1854		1893		1981		1951
Ottawa	1866	1948		1955	1874		1957				1967
Queens	1841	1893		1903	1894		1894	1918	1894		1894
Regina	1965	1978									
Royal Military College	1876	1962		1962	1962		1962	1975			1962
Saskatchewan	1907	1912	1912	1931	1912		1947	1935	1935		1924
Sherbrooke	1954	1954		1971	1954		1954				1954
Tech. Univ. of Nova Scotia	1907	1907	1972	1949	1909		1910		1967	1967	1910
Toronto	1827	1873		1902	1878		1887		1935	1944	1887
Trois Rivières	1969	1969					1969			1973	
Waterloo	1957	1957		1957	1957		1957		1981		1957
Western	1878	1954		1954	1954		1954				1954
Windsor	1957	1957		1958	1958		1958		1972	1972	1958

practical use of instruments, and drawing. The programme was conducted by the University College, which comprised at that time only the Faculty of Arts. Because of the lack of a qualified civil engineering instructor, the applied subjects had to be dropped from the curriculum and the students had to rely on textbooks and such instruction as they could obtain outside the university. However, to orient the course in the direction of civil engineering, notable practising engineers such as Sandford Fleming, Thomas C. Keefer and John Galbraith were appointed as external examiners.

Judged by the number of students, the civil engineering programme could hardly be considered a success. While four people entered it in 1859, in many of the ensuing years none did. In the twenty-six years that it existed, there were only seven graduates; the last in 1878. The first of

Metallurgical	Mining	Petroleum	Mineral	Materials	Surveying	Shipbuilding	Systems	Building	Eng. Science	Forest	
1915	1915	1948	1973								Alberta
1948	1915		1945								British Columbia
			1979		1979						Calgary
											Carleton
											Chicoutimi
								1980			Concordia
1958	1958										École Polytechnique
											Guelph
											Lakehead
1960	1960										Laurentian
1940	1939										Laval
1874	1874										McGill
1957											McMaster
											Manitoba
						1979					Memorial
											Moncton
					1960					1968	New Brunswick
											Ottawa
1903	1894										Queens
							1978				Regina
				1982							Royal Military College
	1962										Saskatchewan
											Sherbrooke
1949	1909										Tech. Univ. of Nova Scotia
1912									1935		Toronto
											Trois Rivières
							1969				Waterloo
				1964							Western
				1965							Windsor

these, C. F. G. Robertson, received the two-year diploma in 1861, thus becoming the first graduate of an engineering programme in Canada.

It became evident that the technological needs of Ontario were not being met adequately by the University College course and in 1869 an appeal was made to the legislature to establish a School of Mines and Mining Engineering in connection with the university. However, the government denied the request and the existing programme was maintained until 1884, although no students were enrolled after 1878.

However, in 1877, the legislature approved the relocation of the School of Practical Science—an artisans' school that had been established in Toronto years earlier—to the University of Toronto Campus, and its transformation to professional status. Almost immediately the university set about to provide facilities and staff. Construction of a three-storey building was begun in the spring of 1877 and completed in 1878. The building provided space for practical work in assaying, mining and physics, and for chemistry laboratories. Interestingly, while there were an engineering lecture room, a drafting room, an office for the Professor of Engineering and an engineering models area, there were no engineering laboratories. On 28 September, the government approved the appointment to the Chair of Civil Engineering of John Galbraith, an outstanding student in the Faculty of Arts civil engineering programme who had become articled to a prominent railway engineer and land surveyor and had gained ten years of valuable experience in civil and mechanical engineering.

Classes in the school began in the autumn of 1878. The programme was of three years' duration, leading to a diploma entitling the graduate to the standing of an 'Associate of the School'. (In 1884, the University Senate replaced the diploma by the degree of Civil Engineer.) Following two common years, the student entered a programme in one of three departments: Engineering (Civil, Mechanical and Mining), Assaying and Mining Geology, or Analytical and Applied Chemistry. Seven full-time students—three in mechanical engineering and four in civil engineering—enrolled for the 1878/79 session.

During the first seven years of the programme, Professor Galbraith had to single-handedly teach all of the engineering courses and his workload was astonishing by today's standards. For example, in 1883/84 he taught fourteen courses of lectures, twelve of them extending through both terms of the session, and in addition he gave practical instruction in surveying, astronomy and drawing, including structural and mechanical design, mapping and topography. In addition to his teaching duties, Galbraith performed all of the duties of a dean and a registrar in the school. Not surprisingly, by 1886, limitations on staff and facilities forced the school to limit its course to civil and mining engineering, although some mechanical and electrical engineering courses were available. On 16 April 1889, a division of Mechanical (including electrical) Engineering was established. When the 1889/90 session opened, there were seventy students either full-time or part-time in the two engineering programmes as well as four in analytical and applied chemistry and thirty in assaying and mining geology.

In November 1889, the school was reorganized and John Galbraith was named Principal and Chairman of a governing council composed exclusively of teachers (which were now eight). Under Galbraith's leadership, development proceeded steadily. With the appointment of an appropriate staff member, a programme in architecture was implemented in 1890. With the completion of an addition to the school's building, a wealth of new equipment, including testing machines, pumps, turbines, tanks, an experimental steam plant and electrical engineering equipment, was installed. By the 1892/93 session, entrance requirements for the engineering programmes had been strengthened and a fourth 'postgraduate' year had been added.

During the succeeding years, there was to be intense competition between the school and Queen's University in Kingston for government support to establish a pre-eminent provincial School of Mines. In the long run, mining schools were developed at both Queen's and Toronto. The school at Queen's was built up primarily through individual and corporate donations, while the mining programme in Toronto eventually received adequate funding from the Ontario Government.

The decade of the 1890s was one of continued growth. The academic staff increased from 7 to 17 and additional equipment was acquired. The student population increased from 90 in 1890 to 230 in 1900, primarily as a result of a surge in the popularity of mechanical and electrical engineering. By 1895, the two programmes claimed 57 per cent of the students. On the other hand, civil engineering, which had earlier attracted most of the students, had declined in popularity and then claimed only 23 per cent.

The place of engineering within the University of Toronto was not established officially until 1906. The School of Practical Science, while located on the university campus, was funded by, and was responsible directly to, the provincial legislature. On the other hand, it shared facilities, courses and staff with the university. The relationship of the school and its staff to the university was somewhat ambiguous and this inevitably led to friction. The ambiguity was finally removed by the University Act of 1906 which reconstituted the school as the Faculty of Applied Science and Engineering of the University of Toronto. Appropriately, the first Dean of the faculty was the man who had laboured so many years in the service of engineering education in Ontario, John Galbraith.

New Brunswick

If the University of Toronto was the first to plan for formal engineering instruction in Canada, the University of New Brunswick was the first to implement such instruction. As was not uncommon in the mid-nineteenth century, there was resistance to the introduction of 'practical' teaching in King's College, Fredericton (which later became the University of New Brunswick). In 1851, the President of the college proclaimed his belief that intellectual or moral culture could not be attained in practical courses:

> To those who would make the college a polytechnic institution we may not promise much more in the way of merely practical teaching; we must not listen to the cry which calls us from the pursuit of truth and virtue to the lower paths and grosser occupations of the multitude.

Nevertheless, despite hostility in the college, the community and the provincial legislature, two men struggled successfully to lay the foundations for engineering instruction in New Brunswick. They were Sir Edmund Head, Governor of New Brunswick, and Dr William Brydone-Jack.

Dr Brydone-Jack came from Scotland in 1840 as Professor of Mathematics, Natural Philosophy and Astronomy. Since the early 1840s he had been giving lectures as part of the mathematics course and had introduced field-work in surveying and astronomy. In 1852, at the urging of Governor Head, the College Council established a budget of £150 per annum 'to defray the expense of lectures and Practical Instruction to be given in Civil Engineering and Drawing'.

The following year, a course of instruction in civil engineering was organized. It included instruction in the use of logarithms, sines, tangents, etc.; resolution of plane triangles; methods of surveying; use and adjustment of instruments; determination of the best route for railways; computation of quantities of land, earthwork, etc; railway curves and gradients; composition and resolution of forces; strength of materials; and theory and practice of timber and iron framing of viaducts and bridges. The instructor, Mr McMahon Cregan, was an engineer brought from England to conduct a survey for the European and North American Railway. Dr Brydone-Jack had laid down the mathematical foundation during the autumn months of 1853 and on 15 February 1854, Mr Cregan began the first formal instruction in professional engineering given in an academic institution in Canada. Twenty-six students were enrolled, a number of whom were regular arts students. The course lasted two and a half months—the winter period during which it was impossible for work on the railway survey to be done.

Following the granting of the Charter to the University of New Brunswick in 1859, courses in engineering were given regularly, even though there was as yet no full-time engineering instructor. Successful students received a certificate. The first such certificate was conferred upon Henry George Clopper Ketchum in June 1862. He subsequently followed a successful professional engineering career and in his will established the Ketchum Silver Medal, which is awarded annually to a top engineering student at the University of New Brunswick.

For many years, the number of students in the engineering classes was small; fewer than five or six diplomas were awarded annually. Due in part to very limited funding, little change occurred in the engineering offering until 1890. In August of 1889, the Senate approved the establishment of a Chair of Civil Engineering, to which Allen Strong was appointed in 1890. In 1893, George Downing, an electrical engineering graduate of the Polytechnic Institute of Boston, was appointed as Professor of Physics and Electrical Engineering. This was the first recognition of electrical engineering in the university.

Although by 1893 chairs had been established in both civil and electrical engineering, graduates of the engineering programmes were still awarded only a diploma, rather than a degree. Indeed, there was still a reluctance within the university to admit the value of technical training. It was only after the engineering students petitioned the Senate, in April

1899, that that body sought and received the approval of the legislature to confer degrees in engineering. In the meantime, plans were under way for the university's Centennial Celebration in 1900. Again, in response to a student petition, the Senate undertook the construction of an Engineering Building. The construction contract, amounting to $15,470 was awarded on 25 May 1900. The Departments of Civil Engineering and Drawing, Physics and Electrical Engineering, and Chemistry took possession of the building in 1901.

With the new facilities for engineering instruction, engineering student numbers increased rapidly in the early 1900s. In 1907, the degree designations were changed to B.Sc. in Civil Engineering and B.Sc. in Electrical Engineering, and an M.Sc. degree was established. In 1908 John Stevens was appointed as the first Professor of Mechanical Engineering, although a mechanical engineering degree programme was not established until 1951.

Nova Scotia

The first school of engineering in Nova Scotia was opened in St Francis Xavier University in 1899. Hugh MacPherson, a native son who had studied engineering at the University of Lille, taught courses in civil and mechanical engineering. Shortly thereafter, in 1907, the Nova Scotia legislature passed an act establishing the Nova Scotia Technical College in Halifax. A site was provided by the Dominion Department of Militia and Defense, on the condition that the college would include military training as a required subject in all departments of engineering. The college was opened in the autumn of 1909, with courses in civil, electrical, mechanical and mining engineering. Its first Principal was Frederick H. Saxton, an engineer educated at the Massachusetts Institute of Technology.

As engineering degree programmes were being developed, it was decided that the college would adopt a novel approach to best satisfy the educational needs of Nova Scotia and its neighbouring provinces. Only the final two years of the programmes were offered, with the preparatory programme years left to 'feeder colleges'. Despite the fact that the college has been restructured four times since 1907, the feeder college system remains in place. Thus, graduates of two-year engineering programmes in Acadia, Dalhousie, Mount Allison, Prince Edward Island, St Francis Xavier and St Mary's Universities transfer to the college to complete their engineering degree programmes. In 1980, the name of the college was changed to the Technical University of Nova Scotia.

McGill University

It was primarily through the inspiration of John William Dawson that formal engineering instruction was established in the Province of Quebec.

In 1855, the Governor-General of Canada, Sir Edmund Head, took the unprecedented step of appointing a scientist, Dawson, rather than a clergyman, as Principal of McGill University in Montreal.

Dawson set out to make McGill a 'contemporary' university with emphasis on the technological needs of a sparsely populated country endowed with few cultural facilities, but with vast natural resources. While many of the early faculty, including Dawson and the first Dean of Engineering, Henry Bovey, were recruited from Great Britain, Dawson did not follow the British university model. While cognizant of the success in engineering education of the *école polytechnique* model in France and the German equivalent, the *technische Hochschule*, he favoured instead the model of the land-grant universities in the United States. The latter combined the polytechnic institute concept with that of the university, but gave primacy to the institute concept. Thus, both liberal arts and science courses were taught, but the primary objective was 'to teach such branches of learning as are related to agriculture and the mechanical arts'. Despite chronic financial difficulties in the early years, Dawson's strategy paid off and by the end of the nineteenth century McGill had attained an international reputation.

Within days of his inauguration as Principal of McGill, Dawson announced a course of thirty popular lectures in scientific subjects, including civil engineering. The following year, 1856, saw the establishment of a two-year civil engineering course in the Faculty of Arts. The first diploma in civil engineering was conferred in 1858. However, government funding of the university was pitifully small, only one quarter of that enjoyed by the University of Toronto, and by 1863 McGill was accumulating an annual deficit of $10,000. Alarmed by the immense deficit, the Board of Governors discontinued the engineering course.

But Dawson was undaunted and continued to promote the revival of the Chair of Civil Engineering. The opportunity arose in 1871 when the Ontario Government announced plans for the establishment of a School of Practical Science at the University of Toronto. Dawson argued that excellent students from McGill would be attracted by the new programme in Toronto, with detrimental effects on McGill. He pointed out that all arrangements were in place for a School of Engineering at McGill and he obtained permission to seek private donations to support it. Three months later, Dawson was able to report that $12,000 had been raised in endowments and annual subscriptions. Thus a rejuvenated Department of Practical Science was inaugurated in the 1871/72 session. Principal Dawson's financial problems were not over, however.

The spectacular rise of the new department caused apprehension and perhaps some jealousy among professors in the older disciplines. They found their prestige threatened and feared that their penurious resources might be further eroded to satisfy the technological needs of the new programme. Moreover, with Confederation in 1867, funding of education had become a provincial responsibility and it was apparent that McGill, an English protestant university, would receive very meagre government support. Admittedly, the Quebec Government did make available to McGill about $1,000 per year derived from marriage licence fees paid by protestants. But as far as the Ministry of Public Instruction was concerned, any subsidies to be allocated specifically for education in applied science in Montreal were to be given to a new 'École Polytechnique' for the training of the French-speaking population.

The largesse of one man, however, subsequently enabled the Faculty of Applied Science of McGill University to establish a reputation as 'the most perfectly equipped in the world'. Sir William Macdonald, a native of Prince Edward Island, endowed most of the professorships, financed most of the building and paid for most of the equipment. He donated more than 13 million dollars to McGill University, though the exact amount is unknown because of the unassuming manner in which he handed out his gifts. Minutes of the Board of Governors contain repeated references to lengthy discussions about financial obstacles, which terminate with the remark that as the meeting was to adjourn Sir William handed a cheque to the Secretary, to be used by the Principal. For several years he contributed $10,000 a year 'to make up any deficiency that might occur in the income of the Faculty of Applied Science'.

In 1877, McGill's Board of Governors decided to convert the Department of Practical Science into the Faculty of Applied Science. Henry Bovey, a Fellow of Queen's College, Cambridge and Assistant Engineer on the Mersey docks, was appointed Dean, with a salary of $1,750 per year. By 1890 the total number of students had risen to seventy-five. Thanks to the munificence of Macdonald and other benefactors (including Thomas Workman, whose bequest of $117,000 permitted the founding of the Mechanical Engineering Department), spectacular growth occurred during the 1890s. By the end of the nineteenth century, nearly 400 students had graduated from the Departments of Civil, Mechanical, Mining, Electrical, Metallurgical and Chemical Engineering. It would be eighty years before any new engineering programme would be added at McGill.

École Polytechnique

The first formal French-language engineering instruction in Canada took place in what is now the École Polytechnique in Montreal. Following Confederation, the Quebec Minister of Public Instruction entered into negotiations with the Catholic School Commissioners concerning the establishment of classes in applied science and the issuing of engineering diplomas by the newly constructed École des Sciences Appliquées aux Arts et à l'Industrie in Montreal. Although the negotiations were not officially made public until late 1873, a sum of $13,340 had been disbursed for the payment of teachers and the purchase of equipment.

Principal Dawson and his colleagues at McGill University were concerned about the apparent low academic level of engineering instruction that was planned for the new institution. Hence, they petitioned the government to consider providing additional financial support to McGill University to permit it to provide engineering instruction in both French and English. None the less, in 1877 the government passed a Bill establishing the École Polytechnique and placing it under the direct control of the Superintendent of Public Instruction.

In 1887, the École Polytechnique became affiliated with the Faculté des Arts de l'Université Laval in Quebec City. However, twelve years later it was incorporated as an independent institution. Finally, in 1923, it became affiliated with the newly established Université de Montreal. In 1905 the school took possession of a newly constructed building on rue Saint-Denis in Montreal. While the building had to be enlarged in 1907 and on several subsequent occasions, it served the school for half a century.

While the civil engineering programme dates back to 1873, diversification of the engineering offerings did not occur until the École Polytechnique obtained a new charter in 1955 and new buildings on the campus of the Université de Montreal in 1958. At that time, new programmes in chemical, electrical, geological, mechanical, metallurgical and mining engineering and engineering physics were implemented.

Queen's University

Another Canadian university to establish engineering instruction in the nineteenth century was Queen's University in Kingston, Ontario. Queen's had been founded in 1841 by the Synod of the Presbyterian Church in Canada. Funds were provided in part by grants from the Presbyterian Church in Scotland and from the Canadian Government, and in part by subscriptions from friends of the university. The Reverend G. M. Grant became Principal in 1877 and it was during his brilliant twenty-five-year tenure that engineering instruction was begun at Queen's.

Following a period of intense competition with the University of Toronto for a government-funded school of mines, the School of Mining and Agriculture was established in Kingston in 1893. The school was affiliated with Queen's University, but was formed independently in order to qualify for provincial government grants. (These grants could not be obtained by the university because the government would not subsidize Church-affiliated institutions.) Meanwhile, in 1891, the university had established a Faculty of Practical Science and appointed Nathan Dupuis its first Dean. A very close interrelationship soon developed between the Faculty of Practical Science and the School of Mining, with members of the faculty serving also on the staff of the new school. By 1894, programmes were offered in civil, electrical, geological and mechanical engineering and beginning in 1896/97 graduates in the Mining School received the B.Sc. degree. There was also provision for the degree of Mining Engineer to be granted on proof that three months had been spent working in a mine. In 1903, programmes in chemical and metallurgical engineering were added and by 1905/06 a Master of Science degree was offered. All degrees were awarded by the university rather than the school—further evidence of the close affiliation that existed between the two institutions.

Five regular degree students were enrolled for the first session of the school and a number of others were enrolled in diploma courses. By 1914, the enrolment had increased to 246 and the faculty staff in the engineering programmes numbered twenty-nine. Two years earlier, the university had separated from the Church, thus making it possible for the School of Mining to become a faculty of Queen's University. Amalgamation was accomplished in 1916, with the establishment of the Faculty of Applied Science.

Royal Military College of Canada

Canada's only degree-granting military college was established in 1874, when the Canadian Parliament passed an Act providing for an institution 'for the purpose of imparting a complete education in all branches of military tactics, fortification, engineering and general scientific knowledge in subjects connected with and necessary to a thorough knowledge of the military profession'. Kingston, with its historical military and naval associations, was selected as the site of the proposed college. On 1 June 1876, the Military College of Canada opened its doors to a class of eighteen cadets. Two years later, Queen Victoria granted the college the right to add the prefix 'Royal' to its name.

Beginning with the first graduating class of 1880, a diploma of graduation was awarded on successful completion of the college course. It was not until the Ontario legislature passed the Royal Military College of Canada Degrees Act in 1959, that the college was permitted to award degrees. The first engineering degrees, in chemical, civil, electrical and mechanical engineering were awarded in 1962.

Canada's other two military colleges, Royal Roads, located near Victoria, British Columbia and the Collège Militaire Royale de Saint Jean, in Saint Jean, Quebec, were established in 1942 and 1952 respectively. The two offer the first two years of the engineering programmes, which are then completed at the Royal Military College of Canada. The latter is one of the very few institutions that offer engineering instruction in both English and French.

Manitoba, Saskatchewan and Alberta

Engineering instruction in Canada's prairie provinces commenced shortly after Saskatchewan and Alberta achieved provincial status in 1905. While the University of Manitoba was established in 1877, it was not until 1907 that the provincial government raised the university's annual grant from $9,000 to $15,000 on the understanding that departments of instruction in civil and electrical engineering would be established. That year, E. E. Brydone-Jack was appointed as Professor of Civil Engineering and the first class of engineering students was enrolled in the autumn of 1907. For two years, Brydone-Jack was the only engineering instructor. Then, in 1909, E. P. Featherstonhaugh was appointed Professor of Electrical Engineering. Because the first two years of the four-year civil and electrical programmes were common, Featherstonhaugh had to instruct only the third and fourth years of the electrical programme. Hence the first graduating class, that of 1911, included seven civil engineers and two electrical engineers. Plans to construct an Engineering Building and to establish a Department of Mechanical Engineering were aborted, with the outbreak of the First World War in 1914. It was not until 1921 that the university established the Faculty of Engineering, appointing Dr Featherstonhaugh as its first Dean.

In Saskatchewan, the new Provincial Legislature passed the University of Saskatchewan Act on 3 April 1907. As with the Universities of Manitoba and Alberta, the University of Saskatchewan, being the only university in the province, quickly implemented instruction in diverse fields. Thus, the College of Engineering was established in 1912. That year, civil and agricultural engineering programmes were begun. However, the impact of the First World War and the great depression was to delay the implementation of mechanical engineering until 1924 and of other programmes well into the 1930s.

In its first session, that of 1906, the Alberta Legislature established the University of Alberta, in Edmonton. Classes opened in 1908 in what is now a public school building. There were forty-five students and a faculty of five. The Faculty of Applied Science (in 1948, renamed the Faculty of Engineering) was formed in 1913. Classes in civil engineering had been initiated in 1908 and although the growth of the university was curtailed with the outbreak of war in 1914, programmes in electrical, metallurgical and mining engineering were begun in 1915. As with the universities in neighbouring provinces, the impact of the great depression of the 1930s was to bring the University of Alberta to a standstill and the development of new programmes was delayed for years.

British Columbia

Though older and more populous than either Alberta or Saskatchewan, British Columbia did not establish its first university until 1915. This late start was a consequence partly of the competition between the province's two principal cities, Vancouver and Victoria, and partly of the existence in the province of a university college that was affiliated with McGill University.

The creation of a British Columbia university had first been advocated in 1877 and in 1890 the provincial legislature passed an act establishing the University of British Columbia. However, the venture failed because of the lack of a quorum at the first meeting of the Senate. Then in 1908, the legislature repealed the earlier Act and passed a new one under which the University of British Columbia operated until 1963. Notwithstanding its incorporation in 1908, the university did not admit its first students until 1915.

Like the universities of Alberta and Saskatchewan, the University of British Columbia was patterned after the land-grant universities in the United States. Its emphasis was on professional education and the extension of the university into the hinterland. Engineering instruction began in 1915, with programmes in chemical, mining, and civil engineering.

DEVELOPMENTS IN THE TWENTIETH CENTURY

The evolution of engineering education in Canada in the twentieth century has been punctuated by two world wars, a worldwide depression, a post-Second World War 'baby boom' and an explosion in scientific and technological development heralded by the launching of Sputnik I.

By the outbreak of the First World War, engineering was being taught in the four western provincial universities, in Nova Scotia, in New Brunswick and in two universities each in Ontario and Quebec. Almost immediately, large numbers of engineering students and professors began enlisting in the armed forces. However, while the scale of engineering instruction diminished, most of the programmes continued to operate. In several of the universities, space was turned over to the armed forces. For example, at the University of Toronto, two-thirds of the students and many of the staff had joined the Canadian Officer's Training Corps when it took over the Mining Building in 1915. The next year, a large part of the Engineering Building was put at the disposal of the Royal Flying Corps.

With the cessation of hostilities in 1918, engineering student and staff numbers swelled significantly. In most of the universities, special accelerated programmes were implemented for the benefit of returning veterans. In addition, rapid industrial development in the 1920s, particularly in Ontario, increased the demand for engineers and thus for places in engineering faculties. Nevertheless, while student numbers grew, most engineering faculties remained primarily undergraduate institutions, with little emphasis on graduate study and research. Indeed, prior to the Second World War, only McGill University and the University of Toronto had acquired international reputations for their engineering research.

With the stock market collapse in 1929, business, manufacturing and construction activity was sharply curtailed. While the country's ten engineering faculties were maintained, student numbers again declined and only three new engineering programmes were introduced in all of Canada during the early and mid-1930s. Ironically, those were introduced at the University of Saskatchewan, a province that suffered grievously from the combined effects of the economic depression and the worst drought ever experienced in Canada.

In 1937, Université Laval in Quebec City became the eleventh Canadian university to offer engineering degrees. Prior to that time, students completed two years of engineering study at Laval and then transferred to another institution, usually the École Polytechnique, to complete the degree requirements. As the École Polytechnique offered only civil engineering at the time, a need was felt to have French-language instruction available in other engineering fields. Consequently, mining engineering was implemented at Laval in 1939 and during the next decade, six additional engineering programmes were introduced.

With the outbreak of the Second World War in 1939, the nation's universities again were placed on a wartime footing. Again, large numbers of engineering students and professors enlisted in the armed forces and again university facilities were converted to military use. More significantly, the Federal Government recognized for the first time that the research competence and technological skills then existent in the universities were valuable assets to be used in the prosecution of the war. Engineering research projects with military applications were undertaken in many universities and while enrolments were reduced, there was feverish activity throughout the war years. Thanks to their successful response to the research challenge, by the end of the war the universities enjoyed a new level of esteem.

Following the war the universities were handed a new challenge, that of undertaking the education of thousands of returning veterans. Under an imaginative programme, the Federal Government provided financial support for veterans to complete university programmes and many veterans opted for engineering. Their impact on 'the system' is illustrated in Figure 2, which shows the numbers of engineering graduates from 1942 to 1983. The number of graduates increased from 746 in 1942 to 3,591 in 1950, the peak year for graduating veterans. The veterans' efforts were to have a second impact on the education system twenty to twenty-five years later. Their children, products of the post-war 'baby boom', swelled the ranks of university graduates in the early 1970s.

The country's eleven engineering faculties performed heroically in accommodating the huge classes of veterans. Professors carried extremely heavy teaching loads and taught classes of several hundred students, while operating under adverse conditions. For example, at the University of Manitoba, engineering classes were conducted in an ancient, poorly insulated ice skating rink. Classrooms were separated by temporary partitions, but not provided with ceilings. And because the building was

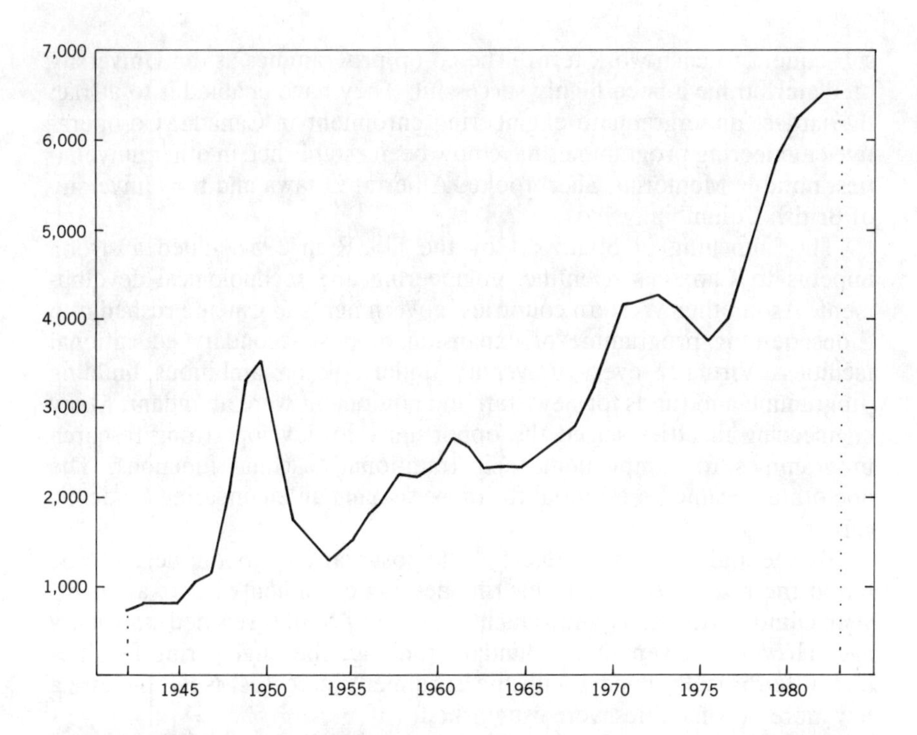

FIG. 2. Engineering graduates' 1942–83.

cohabited by numerous pigeons, conditions were not merely adverse, they were hazardous.

Interestingly, it was not until after the veterans had passed through the university system and student numbers had dropped back almost to the immediate post-war levels that the number of engineering faculties began to expand rapidly. Following an unprecedented post-war period of economic and industrial expansion in Ontario, the provincial government approved the establishment of engineering programmes in the University of Western Ontario, London (1954), the University of Ottawa (1955), Carleton University, Ottawa and the University of Waterloo (1957), the University of Windsor (1958), McMaster University, Hamilton (1959) and Laurentian University, Sudbury (1960).

The University of Waterloo offered the first co-operative engineering programmes in Canada in 1957. The first classes to be offered were those leading to degrees in chemical, civil, electrical and mechanical engineering. Co-op programmes provide an integrated pattern of academic study and industrial experience in which six four-month work terms in an industrial setting are interspersed among eight four-month academic terms spent on campus. Through its Department of Co-ordination and Placement, the university finds employers for the students for the work terms. Both the student and the employer must submit a report subsequent to each work term. The co-op programmes at the University of Waterloo have been highly successful. They have enabled it to attract the largest undergraduate engineering enrolment in Canada. Co-operative engineering programmes have now been established in other universities, notably Memorial, Sherbrooke, Alberta, Ottawa and the University of British Columbia.

The launching of Sputnik I by the USSR in 1957 added a strong impetus to Canada's scientific, engineering and technological development. As in other Western countries, governments in Canada rushed into a decade-long programme of expansion of post-secondary educational facilities. Virtually every university undertook an ambitious building programme and funds for new staff and equipment were abundant. Many engineering faculties seized the opportunity to develop strong research programmes to complement their traditional teaching functions. The doctorate became an essential for those seeking an engineering academic career.

By the mid-1960s, the effects of the post-war baby boom began to be felt at the post-secondary level. Engineering enrolments began an eight-year climb as the sons and daughters of servicemen reached university age. However, given their abundant funding, the engineering faculties had little difficulty coping with the enrolment increases. Nevertheless, a few more new faculties were established.

In 1964, the University of Guelph was incorporated, following the amalgamation of the Ontario Agricultural College, the Ontario Veterinary College and the MacDonald Institute. It began offering degree programmes in agricultural, biological and water resources engineering in 1969.

Sir George Williams University in Montreal initiated engineering instruction in 1963 and then established degree programmes in three areas of specialization in 1966. In 1974, the Quebec legislative assembly ratified the merger of Sir George Williams University and Loyola College and the consequent formation of Concordia University.

In order to satisfy regional needs in the Province of Quebec, the University of Quebec, in 1969, established engineering faculties on its campuses in Chicoutimi and Trois Rivières. The former now offers a general programme, *génie unifié*, whilst at Trois Rivières, the initial programmes were *génie électrique* and *génie industriel*.

In New Brunswick, there was similar motivation for the establishment of programmes in *génie civil* and *génie industriel* at the Université de Moncton, in 1968 and 1972 respectively. A need was felt among the Acadian population of northern New Brunswick for French-language instruction in specialties that would accelerate the industrial development of the region.

While the only engineering faculty in Newfoundland was not established until 1969, its parent institution had its origins in 1925 when a college was established as a memorial to those who had lost their lives in the First World War. Then on 13 August 1949, the first session of Newfoundland's first provincial government established the Memorial University of Newfoundland. From its inception, the Engineering Faculty catered for the needs of the province for expertise in the traditional areas of civil, electrical and mechanical engineering. However, it also recognized the unique needs of Newfoundland as a cold and maritime province. Thus it was instrumental in the establishment, in 1975, of the Centre for Cold Ocean Resources Engineering in Memorial University. In addition, the faculty established Canada's first programme in shipbuilding engineering in 1979.

The School of Engineering in Lakehead University in Thunder Bay, Ontario is unique in Canada in that it offers both technology and engineering programmes. Engineering technology programmes were established in the Lakehead Technical Institute in 1948, to promote technological development in north-western Ontario. The institute became Lakehead University in 1965 and seven years later programmes in chemical, civil, electrical and mechanical engineering were established. The unique feature of the programmes is that they are specifically designed to permit graduates of the technology programmes of Lakehead University or other institutions to progress to engineering degrees at Lakehead. The graduate of a two-year technology programme completes an additional two-year programme in order to qualify for the engineering degree.

For many years the four western Canadian provinces each had a single university. When the need was felt for a second university, there was a tendency initially to establish a satellite campus of the provincial university. Thus, for example, in Alberta, the University of Calgary had its origin in 1945 with the establishment of a Calgary branch of the Faculty of Education of the University of Alberta. During the next twenty years, branches of various University of Alberta faculties were established in Calgary, some of them offering the complete degree programme, others offering only the first year or two. The first year of engineering was offered in 1957 and in 1963 the complete engineering programme was established as a division of the University of Alberta Faculty of Engineering. Finally, in 1964, the Calgary campus gained autonomy as the University of Calgary. The following year, the Faculty of Engineering was formed, with programmes in chemical, civil, electrical and mechanical engineering.

A similar pattern was followed in Saskatchewan. In 1925, Regina College was accepted as an affiliated junior college of the University of Saskatchewan in Saskatoon and in 1934 the United Church of Canada transferred the college and its property to the university. Then in 1959 the university raised the college to degree-granting status and established it as a second campus. A number of faculties and schools were established on the Regina campus and finally, in 1974, the University of Regina was formed as an autonomous institution. A Faculty of Engineering was formed in 1968, but it offered only the first two years of a four-year programme that was completed at the University of Saskatchewan. The faculty initiated a four-year degree programme in systems engineering in 1979.

RECENT TRENDS IN ENGINEERING EDUCATION

During the 1970s and early 1980s several interesting trends have developed in the field of engineering education. In large measure the trends have been dictated or strongly influenced by an economic upheaval that Canada has shared with other Western countries. That upheaval was triggered by rapid and substantial increases in world oil prices initiated by the OPEC countries in 1973.

An early Canadian reaction to the price increases was a sharp upswing in oil and gas exploration, primarily in the Arctic and off the east coast, and the launching of several mega-projects aimed at the extraction of oil from tar sand deposits in western Canada. The two developments provided strong economic stimulus, particularly in Alberta, and a strong demand for engineering manpower, not only in the oil industry but also in the manufacturing and building industries.

The high oil prices of the late 1970s contributed to a world-wide economic recession that was particularly deep and prolonged. In Canada, disagreements between the Canadian and Alberta governments regarding oil pricing, coupled with a world-wide oil glut, caused much of the work on the energy mega-projcets to be curtailed. The consequence was a precipitous drop in economic activity, most acute in Alberta but felt strongly throughout Canada, beginning in 1981. By 1983, more than 8,000 of Canada's 115,000 professional engineers were unemployed. More than a million Canadians were out of work.

These developments influenced engineering education in several ways.

Undergraduate enrolments

During the 1970s constantly increasing numbers of foreign students sought entry to Canada's engineering faculties. Most of those students were from South-east Asia and initially they were admitted to engineering faculties in the most westerly provinces. As the products of the post-war baby boom were then progressing through the system those faculties soon became saturated and the foreign students began to seek entry to universities right across Canada. During the next few years, most engineering faculties began to impose severe limits on foreign undergraduate students through either quotas or differential academic fees.

With the advent of energy-related projects in the mid-1970s and promise of abundant engineering job opportunities, there were large enrolment increases as Canadian students flooded into engineering programmes. Even with the economic downturn numbers continued to swell, as an engineering degree was expected to provide a good opportunity for employment. By the early 1980s most engineering faculties had imposed enrolment limitations and academic entrance requirements were rising each year as the demand for places intensified.

There were also shifts in emphasis among disciplines. Until approximately 1980 civil engineering remained popular because energy-related projects had significant civil content and because many of the foreign students were in civil programmes. However, as oil costs began to impact negatively on Canada's balance of payments federal and provincial governments began to emphasize industrial development in areas such as electronics, communications and computer-aided design and manufacturing—areas with good potential for export earnings. In response to that emphasis enrolments in electrical engineering programmes and in newly established computer engineering programmes rose dramatically leading to enrolment limitations on some of the programmes, while civil engineering enrolments were at a low ebb. Table 2 shows national undergraduate engineering enrolments for the years 1970 to 1983. Undergraduate enrolments by discipline, for 1981/82 are presented in Table 3.

Undergraduate programmes

As illustrated in Figure 3 the number of undergraduate engineering programmes accredited by the Canadian Accreditation Board (almost all programmes are accredited) increased dramatically during the 1960s and early 1970s, the period of unprecedented growth in Canada's universities. Then during the latter half of the 1970s, there was a time of consolidation and ever-tightening university budgets. Hence despite continuous enrol-

TABLE 2. Engineering undergraduate enrolments, 1970–83.

Year	Enrolment	Increase (decrease) over previous year	Percentage increase (decrease)
1970/71	23 552	263	1.1
1971/72	22 645	(907)	(3.9)
1972/73	22 103	(542)	(2.4)
1973/74	21 176	(927)	(4.2)
1975/76	24 842	2 424	10.8
1976/77	26 501	1 659	6.7
1977/78	28 501	2 000	7.5
1978/79	29 446	945	3.3
1979/80	30 758	1 312	4.5
1980/81	32 179	1 412	4.6
1981/82	34 155	1 976	6.1
1982/83	35 869	1 714	5.0

TABLE 3. Undergraduate engineering enrolments by discipline, 1981/82

Programme	Total enrolment	Programme	Total enrolment
Common core programme	8 249	Geological	728
Agricultural	354	Industrial	818
Chemical	2 877	Mechanical	6 364
Civil	4 546	Metallurgical	549
Computer	154	Mining	497
Electrical	6 675	Surveying	172
Engineering physics	539	Other	1 174
Engineering science	459	TOTAL for Canada	34 155

ment increases, few new engineering programmes were developed. Nevertheless there was increasing national pressure for Canada to improve its competitive position in such 'high technology' areas as electronics and computer-aided design and manufacturing. Thus, beginning in 1980, engineering faculties began to create computer engineering programmes, most of which are already oversubscribed and the number is certain to increase. New programmes in industrial engineering and manufacturing engineering were also being developed.

Engineering curricula

As the pace of technological change accelerated during the 1960s and 1970s, not only did the number of different engineering programme

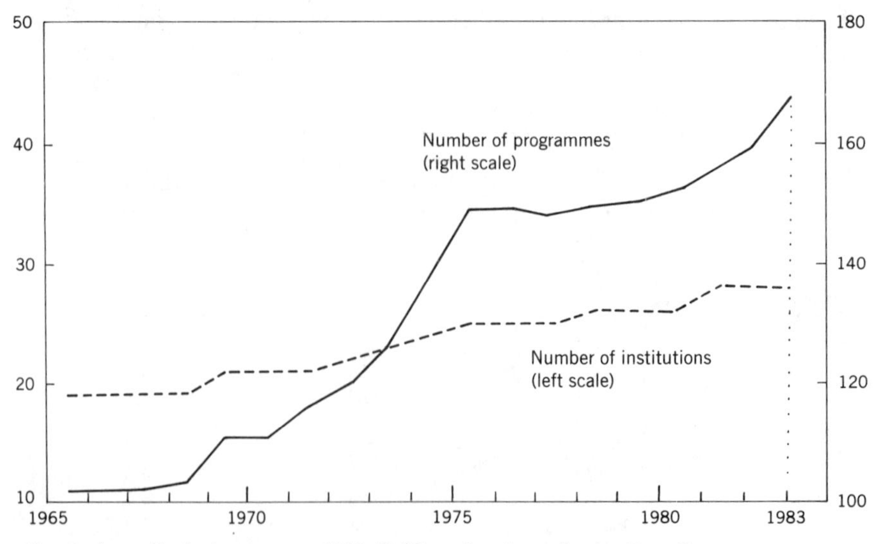

FIG. 3. Accredited programmes, 1965–83 (Canadian Accreditation Board).

offerings increase, the technical content changed and became more concentrated. With the advent of the digital computer all engineering programmes incorporated at least one course in computer methods and programming, and there was an increase in emphasis on numerical techniques amenable to computer applications. By the early 1980s micro-computers were becoming commonplace in engineering organizations and most engineering faculties were incorporating their use in undergraduate and graduate courses.

The strong North American interest in environmental and social issues in the 1960s was reflected in engineering curricula. Due in part to pressure from the Canadian Accreditation Board, engineering faculties increased the content of humanities and social sciences in their curricula to approximately 12 per cent. Then, during the 1970s, many designed courses or incorporated material that would emphasize the role of the engineer in society and the impact of his work on society. At the same time, feedback from the engineering profession suggested that while the technical skills of new engineering graduates were excellent, their written and oral communication skills left much to be desired. In response, some engineering faculties introduced communication courses and others incorporated less formal instruction and practice in communication.

With the knowledge explosion of the 1960s and 1970s, engineering faculties had difficulty in incorporating all of the material that was considered to be essential into the traditional four-year programmes. While a few faculties extended their programmes to five years, by 1984 all engineering programmes outside of Quebec were again of four years' duration. With the establishment of the Collèges d'Enseignement Général et Professionel (CEGEPs) in Quebec, beginning in 1967, the Quebec Government decreed that professional engineering programmes in the province would be no more than three years. The rationale was that the two-year CEGEP programmes (which follow grade 11) plus the three-year engineering programme, could provide an educational experience equivalent to grade 12 plus a four-year engineering programme in other provinces. However, because the engineering faculties in Quebec controlled neither the content nor the quality of the programmes in the province's forty-six CEGEPs, they had great difficulty in complying with the three-year programme requirement. Thus while the programmes may be nominally three years in length, most students require three and a half or four years to complete them.

Graduate enrolments

The precipitous drop in engineering employment opportunities in the early 1980s had a stimulating effect on Canada's engineering graduate schools. During the heady days of the 1960s and early 1970s, the best engineering graduates were much sought after by industrial firms and very few elected to continue their studies at the post-graduate level. However, in the 1980s sharply diminished employment opportunities and greatly improved financial support programmes of the Natural Science and Engineering Research Council, and an increased emphasis on industrially oriented research in universities, combined to make engineering graduate study very attractive. Thus by 1983, a new vitality was evident in engineering graduate programmes as increasing numbers of excellent graduates began to make their presence felt: as illustrated in Table 4, enrolment in engineering programmes at the master's and doctoral level increased from 5,486 in 1981/82 to 6,267 in 1982/83. That increase was the largest annual increase ever recorded. Almost one-third of it was in electrical engineering. Table 5 shows the enrolment distribution by discipline. Approximately one-quarter of the engineering graduate students are in doctoral programmes.

TABLE 4. Engineering graduate enrolments

Year	Enrolment	Increase (decrease) over previous year	Percentage increase (decrease)
1970/71	4 488	216	5.1
1971/72	4 268	(220)	(4.9)
1972/73	3 867	(401)	(9.4)
1973/74	3 914	47	1.2
1974/75	4 539	625	16.0
1975/76	5 194	655	14.4
1976/77	5 189	(5)	(0.1)
1977/78	5 159	(30)	(0.6)
1978/79	5 038	(121)	(2.4)
1979/80	4 930	(108)	(2.2)
1980/81	5 084	154	3.1
1981/82	5 486	402	7.9
1982/83	6 267	781	14.2

TABLE 5. Graduate enrolments by discipline

Discipline	1982/83	Percentage of total	1981/82	Percentage of total
Civil	1 486	23.7	1 297	23.6
Electrical	1 549	25.0	1 297	23.6
Mechanical	913	14.5	779	14.2
Chemical	735	11.7	666	12.1
Metallurgy	285	4.5	262	4.8
Computer	139	2.2	138	2.5
Industrial	102	1.6	102	2.0
Other	1 058	16.8	945	17.2
TOTAL	6 267	100.0	5 486	100.0

University funding

In the face of mounting provincial budget deficits exacerbated by the reduced economic activity of the early 1980s provincial governments throughout Canada were hard pressed to meet the increased funding needs of their universities. Hence, in most provinces engineering faculties faced budgets that were shrinking in real terms while undergraduate and graduate enrolments continued to increase. In order to alleviate the problem, most faculties began to use research income to subsidize their undergraduate instruction. In some engineering programmes the total value of research grants had surpassed the annual operating budget.

Nevertheless, the effect of budget restraint was becoming evident. Short-term accreditation periods for engineering programmes were becoming more common, reflecting the Canadian Accreditation Board's concern regarding the stability of many programmes. Furthermore, criticisms of undergraduate instructional facilities and equipment were becoming stronger and more numerous in accreditation reports.

University–industry co-operation

A positive consequence of the need felt in Canada to improve industrial productivity was a growing spirit of co-operation among government research organizations, engineering faculties and industrial organizations. In the late 1970s and early 1980s a trend began to develop for engineering faculties to emphasize industrially oriented research. At the same time, industrial organizations tended to rely more on engineering faculty members to conduct contract research, and federal and government research organizations began to liaise more closely with university staff. The primary areas of interaction were those that held promise for improving industrial productivity; computer-aided design and manufacturing, electronics, etc.

WOMEN IN ENGINEERING

Traditionally, the engineering profession in Canada has been male-dominated. It was not until 1927 that Elsie Gregory MacGill became the first female graduate in electrical engineering at the University of Toronto, and subsequently the first female member of the Engineering Institute of Canada. She also became the first female aeronautical engineer to graduate from the University of Michigan, in 1929.

Ms MacGill enjoyed a distinguished engineering career. Having flight-tested the first Canadian metal plane, she supervised all engineering work for the Canadair production of the Second World War Hawker Hurricane fighter for the UK and the Curtis Wright Helldiver for the US Navy. Later, she headed her own aeronautical engineering consultancy office.

Despite her achievements, Ms MacGill was chided by newspaper headlines stating that hers was 'no job for a woman'. And despite her courageous example and those of other pioneering women engineers, the number of women in the profession did not begin to increase significantly until the 1970s.

By 1982 the percentage of females among the engineering population had risen to 1.5 per cent. However, today 7 per cent of the current engineering student population is female.

Engineering research

Engineering and scientific research activity was almost non-existent in Canada when the Federal Government took its first initiative to foster and co-ordinate industrially oriented research. On 23 November 1916 the Federal Cabinet appointed a nine-member Honorary Advisory Council and assigned to it the following tasks :

To organize, mobilize and encourage existing research agencies in Canada, so as to utilize waste products, discover new processes and exploit unused natural resources.

To make a comprehensive survey of scientific and industrial research in Canada and to examine the possibilities of extension.

To co-ordinate current research, foster a community of interest among research agencies and link science and technology with labour and capital; and to develop ways and means of enlarging the scope of Canadian research.

The council, renamed the National Research Council (NRC) in 1925, recognized from the outset the importance of the universities in conducting research and in developing a pool of research talent, and so the council voted to endow university scholarships with a portion of its modest funds. It wished also to recommend substantial grants of money to those universities already engaged in research work, but at that time only McGill, Toronto and Queen's would have qualified. The situation in industry was no more promising; only $135,000 per annum was being spent on industrial research in all of Canada.

The university scholarship scheme began rather inauspiciously. In 1916 the council allocated $10,000 but there were not sufficient applications to exhaust even that meagre funding. By 1919, the largest number of scholarships that had been awarded in a year had been eight, six of which had gone to students at McGill University. Matters gradually improved in the 1920s and by the end of its first ten years of operation, the council had awarded 344 scholarships to students in twelve universities.

The university research grant programme was somewhat slower to develop, primarily because of a lack of qualified researchers. In 1919, of the $70,000 that was allocated only $10,000 could be spent. By 1927, only

120 projects had received funding. None the less, by 1928/29 the council's budget had reached $323,000 of which $237,000 was devoted to university research grants and scholarships. Two years later, the National Research Council opened its first research laboratory in Ottawa and accordingly the grants and scholarships funding was cut back to $202,000. Now there were far more applications for both university research grants and scholarships than could be awarded.

As the recession deepened and the council tried to maintain its own research establishment in Ottawa, a recurring casualty was the scholarship programme. At the March 1932 meeting of council, twenty-eight scholarships and fellowships with a total value of $16,000 were awarded for the year. There had been 138 applications. Perhaps nothing demonstrates more vividly the desperate state of the times and the hopelessness of the unemployment problem for young people entering the labour market, than the provision, in 1933, for 'employing' persons in the NRC laboratories without pay. Arrangements were made to place 100 such people.

Then in the late 1930s, with a growing threat of war, the worldwide economic depression slowly began to lift, due in part to gradual rearmament. The need for research as an important component in Canada's expanding national defence effort was expressed ably by NRC President, General A. G. L. McNaughton. His efforts were rewarded as the council's budget was increased from $510,000 in 1936/37 to $900,000 in 1939/40.

By the time the Second World War was declared on 1 September 1939, there had been plenty of advance warning. Astonishingly, Canada was not much better prepared for war in 1939 than it had been in 1914. Thanks to General McNaughton, the National Research Council was more ready for wartime mobilization than were most Canadian institutions, including the military. Practically all of Canada's military hardware was obsolete and the military had no research arm to help in the development of modern equipment. Thus, the council was soon placed on a wartime footing and its normal peacetime activities were relegated to second place. Fortunately, in 1939 the council had a new, well equipped, if sparsely occupied, laboratory building in Ottawa. Furthermore, it was almost immediately approached by nearly every university in the country with the offer to place their facilities at the disposal of the government for work of importance in Canada's war effort. Much of the work that was carried out during the next six years was highly secret and it was carried on without public knowledge.

During the immediate pre-war years and the early years of the war, Britain tended to guard its research secrets and the areas of Canadian military research activity were somewhat restricted. However, as the threat to the British Isles increased, more and more co-operative work was undertaken in Canada. Thus the National Research Council made significant contributions in the areas of radar, uranium fission, aviation medicine, defence against chemical warfare, the development of wooden aircraft, and explosives.

Canada emerged from the Second World War in a very healthy economic position. Thanks to sane and courageous tax policies, most of the war costs had been financed as they occurred. Thus, by 1946/47 the Federal Government was able to implement tax cuts while accumulating annual budget surpluses. Canada's industrial establishment had also grown immensely during the war years. Thus, with industrial activities in many countries reduced or even crippled by the war, Canada was able to sell abroad everything that it could produce.

The National Research Council reflected the national post-war spirit of resurgence and confidence. Not only had the council's staff and facilities grown substantially during the war years, Canada's research community had also been welded into a single co-operative corps. During the years from 1946 to 1951, grants for university research increased from $240,000 to $1,340,000 and post-graduate scholarships increased from $52,000 to $200,000. Furthermore, by the early 1950s 60 per cent of the council's budget went to its engineering divisions.

The 1950s and early 1960s saw unprecedented expansion of university activity, including research and the growth of graduate schools. Contributing factors were the sharp increase in immigration after the war, a more affluent society which enabled more people to attend university, the stimulation of wartime research and the 1957 launching of a satellite by the USSR. During this period, university research funding increased eight-fold but even so the needs of the universities threatened to outstrip the funding.

However, despite the increases in government support, research and development expenditure by Canadian industries lagged behind that of foreign competitors. Many of Canada's firms were subsidiaries of American corporations that conducted their research work in the United States. Many others were small firms with little or no research capability. Consequently, during the 1960s and 1970s Canadian industry faced increasing competition, particularly in 'high technology' areas, from foreign competitors. By the mid-1970s the impact of that competition was being felt in the form of closed factories, Canadian jobs lost and increasing imports of manufactured goods.

In the meantime, several disturbing trends were developing in Canada's engineering educational institutions. While undergraduate engineering enrolments gradually increased during the 1960s and 1970s, buoyed by healthy employment prospects, graduate enrolments remained static. Moreover, the percentage of engineering M.Sc. students who were Canadians and permanent residents decreased from 80 per cent to 64 per cent between 1972 and 1982. At the Ph.D. level, the drop was even more dramatic, from 77 per cent to 45 per cent. Coupled with this was the fact that government support for university research and graduate scholarships was allowed to decline throughout the 1970s. For example, between 1969 and 1979 the decline, in uninflated dollars, was 17 per cent.

Partly in response to concerns expressed by Canada's academic community, the government decided in 1978 to separate the National Research Council's own 'in-house' research function from its programme of postgraduate scholarships and university research grants. The latter programme was assigned to the newly formed Natural Sciences and Engineering Research Council. In 1979, the new council developed a five-year plan that incorporated the following objectives: (a) to increase the supply of highly-trained researchers; (b) to correct the severe obsolescence of research equipment in Canadian universities; (c) to increase the university research effort in areas of identified national concern and; (d) to maintain a strong base of free or discipline-oriented research within the universities.

Under the dynamic leadership of Dr Gordon McNabb, the council increased equipment grants to universities by 132 per cent between 1979 and 1981. It introduced several new programmes to facilitate industrial research by academics and to encourage undergraduates to pursue research careers. From 1979 to 1981, the council's support for its manpower training programmes, including postgraduate scholarships, was increased by 76 per cent. During the same period, grants for targetted research in the 'strategic' areas of biotechnology, communications and computers, energy, environmental toxicology, food and oceans was increased by 61 per cent.

At the mid-point of its five-year programme, the council could claim considerable progress toward the Federal Government's stated objective of increasing its support for research and development from 0.9 per cent of the Gross National Product in 1979 to 1.5 per cent by the mid-1980s. It was anticipated that the objective might be frustrated by a shortage of skilled researchers. Nevertheless, the council's manpower training programme is having a dramatic impact on Canada's engineering faculties. With increasing numbers of excellent students proceeding to graduate programmes and with equipment improvements, a new vitality and interest in research is beginning to be felt.

Technical education

THE EARLY YEARS

Vocational education in Canada dates back to 1668 when Bishop Laval opened trade schools in St Joachim and Quebec City, offering instruction in cabinet-making, carpentry, masonry, roofing and other trades. However, decades of ethnic and religious conflict in Quebec hampered efforts to develop a vocational education system. It was not until 1828 that an Artisans' Institute was founded in Montreal, with a branch in Quebec City. This was the beginning of a polytechnical institution for which the government provided grants. In 1855, practical courses in applied sciences were offered at McGill University in Montreal and in 1878 a Faculty of Applied Sciences and Engineering was organized. In 1897, a provincial act placed thirteen trade schools in Quebec under an Arts and Trades Council and provided for grants to maintain them.

In Ontario, vocational education was commenced at the Ontario Society of Artists' School in Toronto in 1876. Vocational training began in Nova Scotia with the establishment of the Halifax Marine School in 1872. A mining school which opened in Halifax in 1872 was the basis for the Nova Scotia Technical College, now the Technical University of Nova Scotia. In New Brunswick, a Mechanics' Institute, the first of several, was opened in St John in 1839. In 1871 the Common Schools Act in that province provided for public support of the education system, including provision for science instruction.

In the Northwest Territories, which included Alberta and Saskatchewan at the time, an industrial school for Indians was established near Qu'Appelle, Saskatchewan, in 1884. The first programmes included book-keeping and simple accounting. Agriculture was added in 1895. Vocational training in Manitoba was begun in 1900 in Winnipeg's Manual Training Centre.

The Anglican Church in British Columbia sponsored institutes in Fort Hope in 1859 and Barkerville in 1869. In addition, a non-sectarian Mechanic's Institute was established in Victoria in 1864. Government support for vocational education began in 1871 with the passage of an Act Respecting Literary Societies and Mechanics' Institutes.

Vocational education in Canada got its first real impetus through the philanthropic efforts of Sir William MacDonald, who donated much of his wealth to the promotion of education for rural citizens. Through the MacDonald Manual Training Plan, sixteen schools were established and in operation across Canada by 1901. These schools formed the basis for vocational programmes that followed in many provinces.

THE PERIOD 1910–60

Federal Government involvement in technical education began in 1910 with the appointment of the Royal Commission on Industrial Training and Technical Education. In its report in 1913 the commission recommended a ten-year programme of Federal Government expenditure to encourage technical and vocational training. The commission's report paved the way for a succession of federal acts relating to technical and vocational training. Those spanning the years 1912 to 1968 are listed in Figure 4.

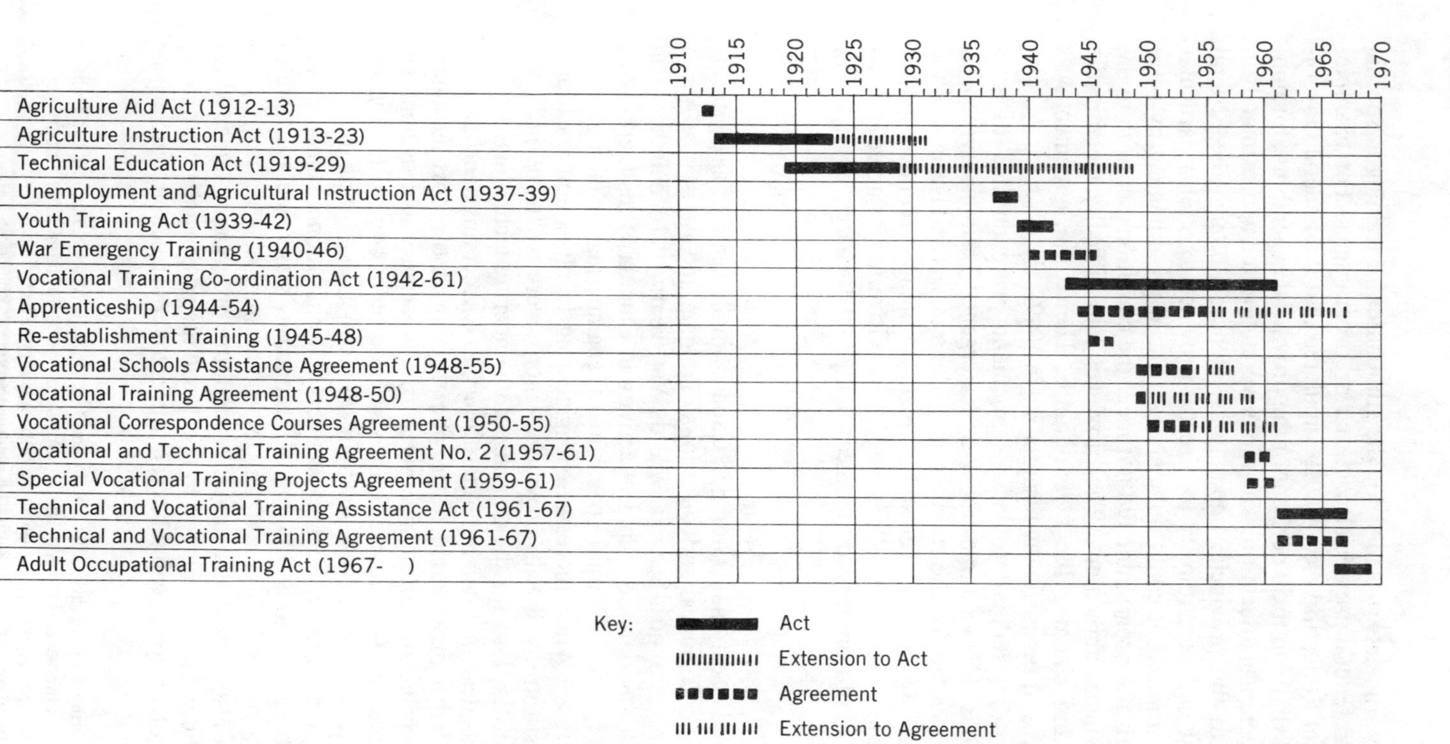

FIG. 4. Federal Acts relating to technical and vocational education.

Typically, the acts provided for the stimulation of training in specific areas, under federal/provincial cost-sharing programmes. For example, the Technical Education Act of 1919 stipulated that the Federal Government would share up to 50 per cent of the provincial expenditure for non-university technical education. The funds were allocated approximately in proportion to the provincial populations. Hence, although the Federal Government had no authority to direct a province to offer a particular type of programme, it could influence provincial programming by withholding funds if a programme did not meet the conditions set forth in the federal/provincial agreement. The provinces had difficulty meeting the conditions and claiming the federal funds. Hence the agreement was extended several times, finally terminating in 1949.

Several other federal/provincial agreements were enacted to meet national needs. For example, the War Emergency Training Agreement (1940–46) provided for the training of workers in war industries and the armed forces. Under the Re-establishment Training agreements signed with all of the provinces in 1945, the Federal Government agreed to pay the entire cost of training programmes for 134,000 discharged servicemen.

POST-1960

Technical and vocational training in Canada finally came of age with the passage of the Technical and Vocational Training and Assistance Act of 1960. During the post-Second World War years there had been an upsurge of industrial and technical activity in Canada. Thus there was a strong demand for a broad spectrum of engineering-technology-vocational expertise. While there were engineering faculties and vocational training institutions in all provinces, there were virtually no post-secondary institutions for the training of engineering technologists.[1] The 1960 Act was designed, among other things, to correct this deficiency.

It included a programme to provide training at the post-secondary level in the principles of engineering, business, science and technology, with emphasis on application. Courses were to be of two or three years' duration with a minimum of 2,400 hours of instruction. During the course of the programme an agreement was reached on national standards for technician training and a Diploma of Technology was made available to each student successfully completing the requirements.

The Act also provided a capital expenditure programme under which the Federal Government agreed to reimburse the provincial governments 75 per cent of approved capital expenditures for construction, purchase or

1. The engineering technologists of Canada and the USA have an education and training which includes theory and practice and equips them for useful immediate employment in industry. They are equivalent to those described as higher technicians (in most countries) or technician-engineers (in the United Kingdom) amongst other designations.

renovation of technical training facilities and for their equipping. By 1970 community college projects valued at $2.16 billion had been approved.

By 1980, there were more than 100 community colleges in Canada. They had been established in every province but Nova Scotia and they ranged in scope from Quebec's forty-six Collèges d'Enseignement Général et Professionel (CEGEPs) covering the whole province, to Newfoundland's single college in Bay St George. Between 1972 and 1979, total enrolment increased from 191,000 to 239,000. The colleges have in common the fact that they are locally oriented, non-degree-granting institutions offering general and specialized programmes for full-time or part-time students. The normal duration of programmes is two years.

With only a few exceptions Canada's engineering technology programmes are taught in the community colleges. However, only about half of these colleges offer engineering technology programmes. As illustrated in Table 6 the large majority of those are located in Ontario and Quebec. The development and current organization of engineering technology institutions vary widely from province to province.

The British Columbia Institute of Technology was established in 1964 to serve the entire province with a variety of engineering technology programmes. Between 1967 and 1977, fourteen regional community colleges were established, primarily to provide continuing education and community service. Four of those include engineering technology or technician courses. Then in 1977 five specialized institutes, one of which offers an engineering technician programme, began operation.

Although Alberta has a network of ten regional community colleges, three religious-affiliated colleges and four vocational centres, engineering technology programmes are offered only at the Northern Alberta Institute of Technology in Edmonton and the Southern Alberta Institute of

TABLE 6. Engineering technology and technician programmes by province

Province	Institutions	Technology programmes	Technician programmes
British Columbia	6	13	3
Alberta	2	40	2
Saskatchewan	2	8	—
Manitoba	2	12	—
Ontario	30	126	114
Quebec	38	167	47
New Brunswick	1	9	—
Nova Scotia	1	3	—
Prince Edward Island	1	3	—
Newfoundland	1	2	—
Territories and Yukon	—	—	—
	84	383	166

Technology in Calgary. Each offers a wide range of two-year engineering technology programmes. Similarly, neighbouring Saskatchewan developed a system of fifteen regional community colleges in the 1970s. None the less, engineering technology programmes are offered only at the Saskatchewan Technical Institute in Moose Jaw and the Kelsey Institute in Saskatoon.

Manitoba's three community colleges were formed in 1969 by restructuring the province's three trade schools. While all three subsequently expanded their programmes and facilities, a range of two-year engineering technology programmes is available only at the Red River Community College in Winnipeg. Only electrical engineering technology is available at the Assiniboine Community College in Brandon.

In Ontario in the late 1960s, a network of Colleges of Applied Arts and Technology (CAATs) was established from five existing institutes of technology and several vocational and trade schools. The CAATs were to offer para-professional career programmes, preparing people to work at a level between the university graduate and the skilled tradesman. At present there are 22 CAATs with more than 90 campuses. Those colleges, together with Lakehead University in Thunder Bay and the Ryerson Polytechnical Institute in Toronto, offer 126 three-year technology programmes and 114 two-year technology programmes. Ryerson offers a wide variety of three-year technology programmes, while Lakehead offers a unique combination of two-year technology programmes and engineering degree programmes.

The CEGEPs are Quebec's response to the community college model. They are comprehensive colleges with two- or three-year post-grade 11 terminal programmes at the technical level and two-year programmes for university entrance. The latter programmes are a requirement for university entrance. Prior to 1964, education in Quebec was largely the domain of the church. However, in that year the first Ministry of Education was founded and in 1967 a Bill was passed establishing the CEGEPs. Within three months, three had been formed and today there are forty-six CEGEPs; four using English and the remainder French. Thirty-eight of the regionally distributed CEGEPs offer more than 200 different two- and three-year engineering technology programmes.

Prior to 1973, post-secondary non-university programmes in New Brunswick were offered through two vocational schools, two institutes of technology and five trade schools. In 1973, an Act of the legislature created the eight-campus New Brunswick Community College, through an amalgamation of those institutions. Today, nine two-year technology programmes are offered on the St John, Moncton and Bathurst campuses. Instruction is in French at Moncton and in English on the other campuses.

While Nova Scotia has a long history of non-university post-secondary education, it has no community college system comparable to those in the other provinces. Rather, the provincial Department of Education operates two vocational centres and three technical institutes; the Nova Scotia Institute of Technology, The Nova Scotia Land Survey Institute and the Nova Scotia Nautical Institute. Three two-year engineering technology programmes, the only ones offered in the province, are at the Nova Scotia Technical Institute in Halifax.

ENROLMENT TRENDS

Engineering technology enrolment figures for all provinces have been available only since 1977. As can be seen from Table 7, there were gradual increases in all provinces between 1977/78 and 1980/81, followed by a 13.5 per cent jump in 1981/82. The distribution by discipline of 1981/82 full-time students is presented in Table 8. The number of engineering technologists graduating each year is now approximately 90 per cent of the number of engineering graduates.

TABLE 7. Engineering technology enrolments by province

Province	1981/82	1980/81	1979/80	1978/79	1977/78
British Columbia	1 962	1 744	1 917	1 927	1 883
Alberta	4 227	3 349	2 919	3 215	2 934
Saskatchewan	515	508	455	429	335
Manitoba	681	636	626	714	663
Ontario	9 932	9 096	8 375	7 976	6 918
Quebec	14 870	12 927	13 250	12 239	11 201
New Brunswick	677	640	608	553	467
Nova Scotia	470	463	452	476	499
Prince Edward Island	136	121	115	111	112
Newfoundland	533	469	278	270	263
TOTAL	34 003	29 953	28 995	27 910	25 275

TABLE 8. Engineering technology enrolments by discipline, 1981/82

	Civil	Chemical	Electrical & Electronics	Industrial	Mechanical	Mining	Biological science	Other	TOTAL
B.C.	564	89	611	0	156	58	484	0	1 962
Alta	1 694	281	924	210	269	403	446	0	4 227
Sask.	183	0	152	0	84	0	96	0	515
Man.	181	37	386	0	55	22	0	0	681
Ont.	3 029	839	3 157	698	1 850	340	19	0	9 932
Que.	3 852	272	6 379	36	2 216	733	796	71	14 355
N.B.	215	52	336	0	74	0	0	0	677
N.S.	103	13	168	0	135	51	0	0	470
P.E.I.	46	40	50	0	0	0	0	0	136
Nfld	106	0	248	0	115	0	64	0	533
TOTAL	9 973	1 623	12 411	944	4 954	1 607	1 905	71	34 488

The engineering technical societies

The first movement toward the organization of an engineering technical society in Canada was spearheaded by Sandford Fleming, the noted surveyor and railway engineer, in the 1860s. However it was not until the Canadian Pacific Railway was under construction in the 1880s and there was engineering activity across Canada, that conditions were right for the formation of an engineering society.

At a meeting of engineers in Montreal on 4 March 1886, a resolution was passed that a society of engineers be formed in Canada, embracing all areas of practice. A committee was appointed to meet with committees of engineers in other cities to arrange for the drafting of a preliminary constitution. Following meetings in Toronto and Ottawa, it was decided that the society should be named the Canadian Society of Civil Engineers. The charter of the new society was carried through Parliament by Walter Shanly, an engineer and Member of Parliament. It received Royal assent on 23 June 1887.

The objects of the society were 'to facilitate the aquirement and interchange of professional knowledge among its members, and more particularly to promote the acquisition of that species of knowledge which has special reference to the profession of civil engineering'. The term 'civil engineering' was intended to refer to all types of engineering other than military. However, while mechanical, mining and electrical engineering disciplines were beginning to develop in Canada, most of the early members of the society were engaged in railway surveys or construction, or in municipal work.

Almost immediately the society set out to establish local branches to promote its activities in various Canadian cities, the first branch being formed in Toronto in 1890. By 1937 there were twenty-five branches and in 1982 there were branches in all of Canada's major cities.

As membership grew and the diversity of activities increased, the society's activities and the branch organization were divided into four sections: general, electrical, mechanical and mining. As a further response to the emergence of a variety of engineering disciplines, on 15 April 1918 an Act of Parliament was passed which changed the name of the society to the Engineering Institute of Canada. That same year the *Engineering Journal* was established as the organ of the institute and membership and technical activity began to increase dramatically.

At the institute's Annual General Meeting in Ottawa in February 1919, a committee was appointed to draft a model Engineering Profession Act which could be recommended to the various provincial legislatures, to provide a framework for the governance of the profession of engineering. The model Act was approved by the membership in July 1919 and in 1920, Acts based upon it were passed in six of the provinces. Subsequently, similar Acts were passed in the other provinces and territories.

In each province the Engineering Profession Act established an Association of Professional Engineers and empowered it to govern the practice of engineering within the province. From the outset, relations between the institute and the provincial associations were complicated by the fact that the former was federally incorporated, while the latter were creatures of the provincial legislatures. Furthermore, while the practising engineer's membership in the provincial association was mandatory, membership in the institute, a learned society, was not. Consequently, while almost all engineers became members of their provincial association, fewer than half of them belonged to the institute.

The provincial associations developed rapidly in the 1920s and by about 1934 they felt the need for a national council. They approached the institute with a request that it provide the necessary framework. In 1935, when agreement could not be reached, the associations established the Dominion Council of Professional Engineers, an organization completely independent of the institute.

In the late 1950s a second attempt was made to bring the institute and the associations together. A joint committee developed a proposal to form a federal body that would cater to both the professional and technical needs of the engineering community. The members of the associations, in a national referendum, voted in favour of the proposal. However, the members of the institute rejected it.

Thereafter the institute entered a period of decline. At the time of the referendum the associations had a total membership of about 35,000, of which about 15,000 were members of the institute. By 1970 association membership had increased to 70,000, while that of the institute had declined to 10,000. It had become apparent that the institute was no longer able to function satisfactorily as a direct membership society catering to the technical needs of members in the diverse fields of engineering.

The mechanical engineers spearheaded a campaign to change the institute by-laws to permit the establishment of constituent technical societies while leaving the institute to provide a co-ordinating role. The Canadian Society for Mechanical Engineering (CSME) was established in 1970, to be followed by the Canadian Geotechnical Society (CGS) and the Canadian Society for Civil Engineering (CSCE) in 1972 and the Canadian Society for Electrical Engineering (CSEE) in 1974.

Despite the heroic efforts of volunteer officers, and a wealth of excellent technical programmes, the Engineering Institute and its technical societies have experienced great difficulty in attracting new members. Membership figures for 1974 and 1983 are presented in Table 9.

The institute and its constituent societies hold national technical conferences and most of them programme a series of continuing education short courses immediately prior to the annual meeting. Most of the constituent societies also hold specialty conferences and continuing

TABLE 9. Membership in the Engineering Institute of Canada and its constituent societies

	1974			1983		
	Practising	Students	Total	Practising	Students	Total
Engineering Institute of Canada, general members	7 360	1 496	8 856	2 188	149	2 337
Canadian Society for Mechanical Engineering	3 080	1 493	4 573	2 473	1 297	3 770
Canadian Society for Civil Engineering	3 190	772	3 962	4 826	1 230	6 056
Canadian Society for Electrical Engineering	—	—	—	1 024	216	1 240
Canadian Geotechnical Society	506	51	557	1 139	27	1 166
TOTAL	14 136	3 812	17 948	11 650	2 919	14 569

education courses from time to time and most of them also maintain local organizations in major cities. Those local sections provide programmes of periodic technical paper presentations, continuing education courses and social events.

The constituent societies maintain co-operative arrangements with sister societies in other countries, holding joint conferences, providing member rates for publications, etc.

The professional engineers associations

Following the First World War there was rapid economic development, particularly in western Canada, which entailed a great deal of engineering construction. Regulations controlling construction were often rather lax and unqualified persons sometimes held positions of engineering responsibility. Although no disastrous failures were attributed to these circumstances, it was evident that waste of materials and faulty design had occurred. Because of the concern felt, committees of the Engineering Institute of Canada in several cities began to develop proposals for legislation which would define engineering, set legal standards of qualifications for its safe practice and give legal status to those who qualified and became registered to practice.

After receiving petitions from all of its branches, the institute arranged to have a model Engineering Profession Act prepared. Following its approval by the membership in 1919, it was used as the basis of engineering profession Acts which were passed in 1920 in British Columbia, Quebec, Alberta, New Brunswick and Nova Scotia. Subsequently, the remaining four provinces and the two territories enacted similar legislation; the most recent to do so being the Northwest Territories in 1979.

Each provincial and territorial Act defines the scope of engineering activity, the administrative structure of the association and the requirements for membership. Members are included on the association's register of professional engineers and are permitted to use the designation 'P. Eng.' (for Professional Engineer). They are provided with an official seal bearing the member's name and the inscription 'Registered Professional Engineer'.

Members are required to seal all engineering documents for which they are responsible. The requirements for membership are appropriate academic preparation and acceptable engineering experience (normally two years) acquired under the supervision of a registered Professional Engineer. The academic requirement may be satisfied in one of two alternative ways: by holding a recognized bachelor's degree in engineering or by completing a programme of technical examinations prescribed by the association. As yet, none of the associations has a programme of compulsory continuing education for its members.

The years of formation of the various associations and their 1982 membership are given in Table 10.

TABLE 10. Provincial associations of professional engineers: dates of formation and memberships

Acronym	Full name	Year of formation	No. of registered members (1982)
APEBC	Association of Professional Engineers of British Columbia	1920	10 970
APEGGA	Association of Professional Engineers, Geologists and Geophysicists of Alberta	1920	18 746
APES	Association of Professional Engineers of Saskatchewan	1930	2 651
APEM	Association of Professional Engineers of Manitoba	1920	2 608
APEO	Association of Professional Engineers of Ontario	1922	50 082
OIQ	Ordre des Ingénieurs du Québec	1920	25 009
APENB	Association of Professional Engineers of New Brunswick	1920	1 712
APENS	Association of Professional Engineers of Nova Scotia	1920	2 356
APEPEI	Association of Professional Engineers of Prince Edward Island	1955	118
APEN	Association of Professional Engineers of Newfoundland	1952	1 094
APEYT	Association of Professional Engineers Yukon Territory	1955	400
NAPEGG	Association of Professional Engineers, Geologists and Geophysicists of the Northwest Territories	1979	90
			115 836

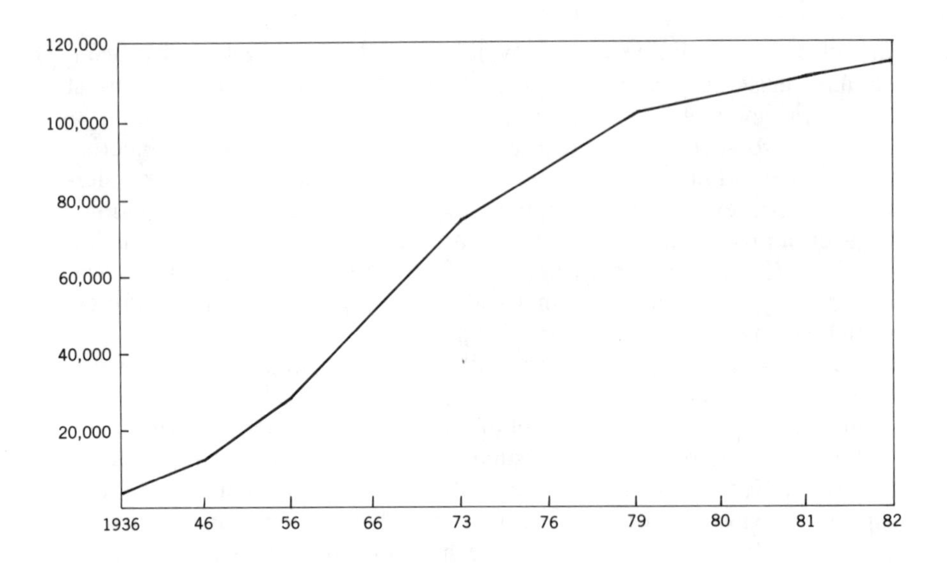

FIG. 5. Registered professional engineers in Canada, 1936–82.

Figure 5 illustrates the growth in the number of registered professional engineers in Canada for the period 1936–82.

The Canadian Council of Professional Engineers

During the period 1926–35, officers of the Engineering Institute of Canada met several times with representatives of the provincial associations to explore ways of consolidating the engineering profession in Canada. Finally, after several inconclusive meetings, representatives of the eight associations then in existence convened in Hamilton, Ontario on 6 February 1936, in conjunction with the Annual Meeting of the institute. At that meeting, the Dominion Council of Professional Engineers was formed and its constitution was drafted. Within two years the constitution had been ratified by the provincial associations in Nova Scotia, New Brunswick, Ontario, Saskatchewan, Alberta, British Columbia, Quebec and Manitoba.

The duties of the council were enunciated as follows:

To assist the provincial associations in securing improved legislation regarding the regulation of professional engineering activity.

To secure the adoption by provincial associations of uniform standards of examination and membership.

To arrange for reciprocal member privileges between provincial associations.

To secure harmony of action in matters of common interest, and to act in an advisory co-ordinating capacity to the provincial associations.

To negotiate with other organizations in advancing the common interests of the engineering profession.

In 1957, the name of the council was changed to the Canadian Council of Professional Engineers and the following year an office was established in Ottawa. Newfoundland had joined the council in 1953 and was followed by Prince Edward Island in 1955. When the Northwest Territories joined in 1980, the council could at last claim to represent every registered Professional Engineer in Canada.

The council decided in 1965 to establish the Canadian Accreditation Board to accredit Canadian engineering programmes. Patterned after the accreditation board of the Engineers Council for Professional Development in the United States, the new board employed similar accreditation criteria and procedures.

Today, the Canadian Council of Professional Engineers is a federation of Canada's ten provincial and two territorial associations of professional engineers that represent Canada's 116,000 registered Professional Engineers. The Board of Directors of the council includes one director from each of the associations. The council is supported by an annual per-capita levy on all association members.

Its activities comprise three facets: accreditation of university undergraduate engineering programmes, data gathering on manpower supply and demand and liaison with governments and with national and international organizations concerned with issues affecting the engineering profession. The council's actions are not binding on its constituent provincial associations. Rather, it provides co-ordination, information and advice to the associations.

The Canadian Accreditation Board

One of the tasks of the association of professional engineers in each of Canada's provinces is to assess the academic qualifications of applicants for admission. Each year, more than 5,000 people graduate from Canada's engineering faculties and several thousand immigrants and potential immigrants apply for academic assessments.

Because of the heavy workload in conducting individual assessments of such large numbers of applicants, the provincial associations began the practice of extending blanket recognition of academic qualifications. Each association considered the graduates of the engineering programmes offered in its province to be qualified academically for admission. Normally this action was taken in the absence of a formal assessment of the programmes.

Over the years, through informal arrangements, the associations extended the blanket recognition to graduates of some of the engineering programmes offered in other provinces. Unfortunately, the recognition was not universal and an engineer who was registered in one province often was required to undertake examinations in order to satisfy the academic requirements in another.

Similarly, several of the associations compiled lists of foreign engineering institutions of strong reputation, whose graduates were assumed to be qualified academically for admission. However, the foreign programmes were not accredited and there was not uniformity among the lists of the various associations.

Recognizing the potential benefits of a co-operative approach to the assessment of engineering academic qualifications, the Canadian Council of Professional Engineers in 1965 established the Canadian Accreditation Board and gave it the following mandate:

To promote uniformity of accreditation of educational qualifications for registration purposes.
To foster, in co-operation with the educational institutions, a high standard of engineering education.
To provide a medium for the interchange of ideas between the universities and the profession regarding the educational needs of the profession.
To assist the associations in the assessment of foreign engineering qualifications.

In carrying out its mandate, the Board engages primarily in three types of activity:

1. It maintains accreditation criteria for engineering programmes in Canada and it conducts regular visits to Canadian engineering faculties to determine whether or not they meet the criteria. It offers constructive criticism and suggestions regarding the programmes.
2. It maintains contact with accreditation bodies in other countries and establishes bilateral mutual recognition agreements where possible. To date, only one such agreement exists: that with the Accreditation Board for Engineering and Technology in the United States.
3. It maintains a set of examination syllabi for assessing the academic qualifications of applicants who are not graduates of accredited engineering programmes.

The board has representation from the various provincial associations of professional engineers and its function is to provide information and advice to them with regard to the academic qualifications of applicants for registration. It meets three times a year.

The primary accreditation criterion is the curriculum content of the engineering programmes. The current requirements for a normal four-year programme are one half-year each of: mathematical foundations; basic science subjects; engineering science subjects; design and synthesis content; and humanities–social science–administrative studies content. That leaves one half-year in which particular emphasis may be incorporated. Other accreditation considerations are: staff qualifications and experience; laboratory and classroom facilities; level of research activity; and computer and library facilities.

The maximum period for which a programme may be accredited is five years. The board, in its *Annual Report*, publishes a list of all currently accredited programmes, with the years in which they were first accredited. The 1983 list includes 175 programmes in 28 institutions. Fewer than ten engineering programmes in Canada are not accredited.

The societies of Certified Engineering Technicians and Technologists

Professional recognition of technologists as an occupational group in Canada dates back to 1927 when the Quebec Government established the Corporation of Applied Sciences Technologists of Quebec. The Act which established the Corporation defined a field of practice for its members and gave them professional status and the exclusive right to the title 'Applied Science Technologist'. However, the corporation was not limited to engineering technologists. It included those in architecture, chemistry, land surveying, agronomy, forestry, engineering evaluation, electricity and tube works.

The first step toward the formation of an organization of engineering technologists and technicians occurred in Ontario in 1957. The Association of Professional Engineers of Ontario, at the instigation of the provincial government, initiated a programme of certification of engineering technologists and technicians. Then in 1962, the Ontario Association of Certified Engineering Technicians and Technologists (OACETT) was registered under the Corporations Act of the Province of Ontario. About the same time, similar societies were established and incorporated in other provinces.

The societies have in common the fact that they are provincially incorporated, non-profit, self-governing bodies dedicated to the maintenance of high standards in the practice of technology. Many of them participate in the development of continuing education programmes for members and have provision for the reclassification of members who enhance their qualifications. Members subscribe to a code of ethics and are entitled to use an appropriate title: Certified Engineering Technologist (CET) in most provinces.

A major interest of the societies has been the definition of the roles of the engineering technologist and the engineering technician and their relationship to other groups, most notably professional engineers. Typically, the requirement for registration as an engineering technologist is graduation from an accredited engineering technology curriculum followed by two years of satisfactory employment experience. That for registration as a technician is graduation from an accredited engineering technician curriculum plus two years of satisfactory experience. The most significant difference between the technology and technician curricula is that the former has a somewhat more extensive mathematical and analytical base than the latter and it is less narrowly specialized. Similarly, the occupational functions of the technician are normally narrower than those of the technologist. Technicians normally receive more detailed direction than do technologists, many of whom advance to some level of management responsibility.

A source of concern to many of the societies of certified engineering technologists and technicians is the degree of professional responsibility that can be assumed by professional engineers and by certified engineering technologists. In most provinces, the right to assume professional responsibility for engineering works is restricted to registered professional engineers. In several provinces attempts have been made to provide engineering technologists with the right to assume professional responsibility for certain types of engineering work. The Engineers Act in Quebec has for many years permitted a technologist or any other person to provide services in connection with non-public buildings of less than a hundred thousand dollars. Quebec Bill 98 passed in 1980 established the Corporation Professionelle des Technologues and an amendment to the Engineers Act emphasized that it does not limit the right of members of the corporation only to perform work for which they have been trained. Technologists in Quebec continue to strive to broaden the area in which they can provide services to the public.

In 1977, the Attorney-General of Ontario appointed a Professional Organizations Committee to review the current Acts governing several professions, including engineering, and to make recommendations for 'comprehensive legislation setting the legal framework' within which the professions would operate. The 1980 report of the committee contained recommendations intended to clarify the engineer/technologist interface. The Society of Engineering Technologists of British Columbia recently proposed an applied sciences technology Act to the provincial legislature. The Act would address the areas of permissible practice of technologists and technicians. Despite the efforts to date to resolve it, the question of the engineer/technologist interface promises to remain a contentious one for some time.

In 1972, the Canadian Council of Engineering Technicians and Technologists (CCETT) was formed as a federation of the ten provincial societies. At present, the council represents approximately 27,000 engineering technicians and technologists, although an estimated 50,000 are practising in Canada. Since 1972, CCETT has been working to strengthen the ties among its members by co-ordinating and facilitating information exchange among them and with similar international organizations. It has produced syllabi of examinations in civil, mechanical, electrical, electronics, chemical and metallurgical engineering technology and plans are in place for syllabi in other discipline areas. In 1982, CCETT formed the Canadian Technology Accreditation Board and by 1983 the board had produced a list of accredited programmes in three provinces, although the Ontario Colleges of Applied Arts and Technology are forbidden by government policy to seek accreditation.

In order to be accredited, the curriculum for an engineering technology programme must be based upon a core of 15 to 20 per cent of applied science and mathematics, including differential equations, statistics, linear algebra and computer science, with application in design and analysis. There should be science and engineering fundamentals applicable to a particular branch of technology complimented by laboratory work amounting to approximately one-third of the total technical curriculum. Finally, approximately 20 per cent of the curriculum should be devoted to liberal studies.

The engineering technician curriculum prescription is similar to the technology one, except that the mathematical content is less extensive and the technical focus is narrower and more empirical.

The year 2000 and beyond

During the remainder of this century, engineering and engineering technology education will be shaped primarily by two factors: a dramatic slowdown in population growth and changing employment opportunities in an increasingly technologically oriented society. Obviously the size of the student population will be dictated in large measure by the size of the cohort of 18- to 22-year-olds each year. It will be affected also by the level of economic activity. The relative prominence of the various engineering disciplines will reflect the changing needs of a society that is becoming increasingly dependent upon technology.

POPULATION TRENDS AND STUDENT NUMBERS

Canada's population growth during the past hundred years has been about 1 to 1.5 per cent per year, except for the periods 1900–20 and 1945–70, when the rate of increase ranged from about 1.7 to 3.0 per cent per year. Approximately 80 per cent of the increase has been due to natural causes (the difference between the birth rate and the death rate) and the remainder to a net migration into Canada. During the late 1940s and the 1950s, the rapid population growth was the result of the post-war baby boom and high immigration rates. However, beginning in about 1960, the birth rate began to decline sharply. Today Canada's birth rate is at an all-time low and it is still declining. Furthermore, as a result of economic adjustments brought on by rapidly increased oil prices, immigration declined throughout the 1970s and early 1980s. A consequence of these phenomena is that Canada's population, which has been ageing for more than a century, has done so at an unprecedented rate since the 1960s. Between 1961 and 1981 the mean age increased from 30.5 to 32 years, while the median age increased from 26.3 to 29.5. During the same period, the percentage of the population aged 0–14 (the pre-working-age group) declined from 34 to 22.5 while the percentage of those 65 and older increased from 7.6 to 9.7.

And what of the future? A number of population projections have been formulated, based upon a variety of assumptions regarding factors such as birth rates and immigration. The projections have been found to be highly consistent and insensitive to changes in the assumptions. Perhaps the most important projection is that between 1981 and 2001 the population will increase from just more than 24 million to about 26.5 million; an increase of less than 1 per cent per year during the 1980s to approximately 0.5 per cent per year in the 1990s. Although at present the number of births per year is increasing slightly (a reflection of post-war baby boom), that number is expected to reach a peak by 1985 and to then decline for several decades.

As the birth rate declines, it is projected that the average age of the population will increase at an accelerating rate. Indeed, the median age, 29.5 years in 1981, is expected to reach approximately 37 years by 2001. This will be a reflection of the decline of the 0–14 age-group from 22.5 to about 19 per cent of the population and the increase in the percentage of those 65 and older from 9.7 to about 12.5 per cent. These trends and projections are illustrated very graphically in Figure 6. Because the 0–14 age-group has been shrinking more rapidly than the over-65 group has been expanding, the proportion of the population (and the absolute

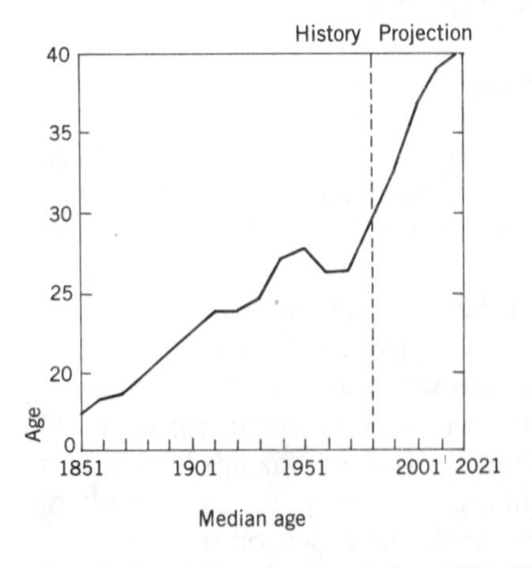

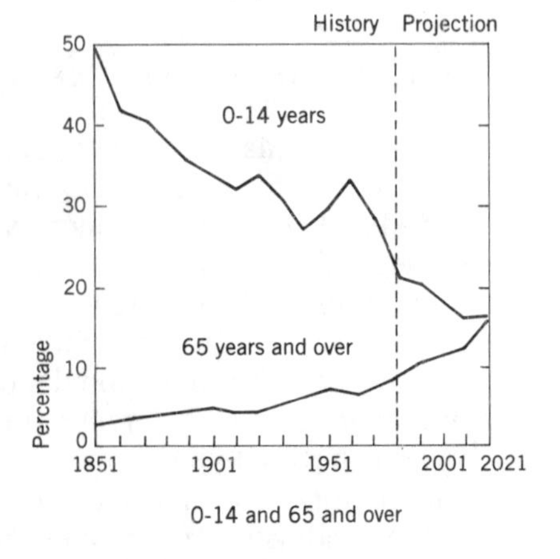

FIG. 6. Canada's ageing population, 1851–2021.

number of people) of working age (15–65 years) is at an all-time high. It is expected to continue to increase, but only modestly.

The recent sharp decline in the number of Canadians in the 0-14 age-group is reflected in the declining number of elementary and secondary-school students. The number of such students (grades 1–12 in most provinces) reached a maximum of 5.84 million in 1970–71. It has declined every year since then and it will drop to approximately 4.8 million in 1984–85. There is likely to be a modest increase, to approximately 5 million by the year 2000, followed by a gradual reduction over ensuing decades.

The post-secondary participation rate (the proportion of high school graduates who enroll in post-secondary institutions) has increased gradually for many years. Nevertheless, it is a virtual certainty that enrolments in both universities and community colleges will follow the trends already experienced in the primary and secondary schools. Thus, Canadian university and non-university (primarily community college) post-secondary enrolments are projected to vary approximately as illustrated in Figure 7. It can be seen that total university enrolment is projected to reach a peak value of 640,000 in about 1984 and then to decline gradually for two decades before the grandchildren of the post-war baby-boom generation pass through the system in about 2010. The projections for the non-university institutions are similar, except that the enrolment peak occurred in about 1981.

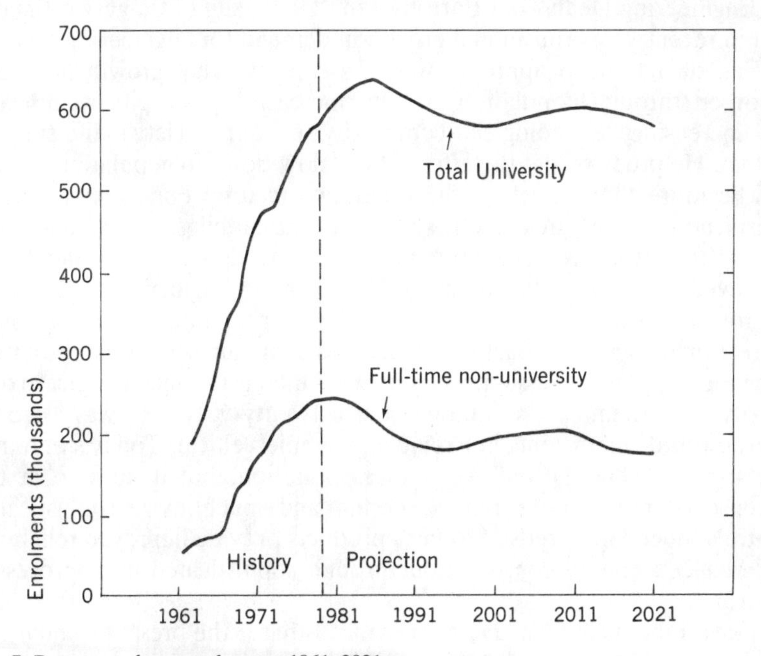

FIG. 7. Post-secondary enrolments, 1961–2021.

ENGINEERING EMPLOYMENT OPPORTUNITIES

During the past several decades, Canada's engineering population has grown at a rate of about 5 per cent per year. The engineering technologist population has grown at a slightly faster rate. During the same period, approximately 3,000 foreign engineers were in Canada each year on work permits, and typically about 1,000 Canadian engineers were 'between jobs' at any given time. At the height of the economic recession in the early 1980s approximately 8,000 Canadian engineers were unemployed. However, by 1984, that number had dropped to approximately 5,000. Traditionally, engineering enrolments have tended to increase in buoyant economic times and to decrease during recessionary periods. However, because of the time-lag of about eight years between the student's career choice (normally made about grade 10) and graduation, it has often happened that large graduating classes coincided with downturns in the economy and small ones coincided with upturns. Consequently, there have been times when engineering and technology graduates were much sought after and other times when many of them could not find immediate employment.

With Canada's population growth projected to be less than 1 per cent per year during the next several decades, it might be expected that economic growth will be modest and that engineering employment prospects will be bleak. However, there are reasons for optimism.

Engineering Dean Tom Barton of the University of Calgary estimates that in recent years the annual growth in demand for engineering services in Canada has been approximately 7 per cent. That growth has been absorbed through population growth, increasing per-capita wealth and the increasing technological complexity of our society's life-support system. He predicts that the effect of the slow-down in population growth will be more than offset by the other two factors. For example, while conspicuous wealth in the form of more automobiles, colour television sets, refrigerators, etc. per capita may not be needed, society will demand improved environmental quality in the form of cleaner air and water, improved waste disposal, lower energy consumption and improved recreational facilities. Barton emphasizes that our society is absolutely dependent upon a life-support system of increasing technological complexity. For example, the automobile of twenty years ago was a crude, overpowered, inefficient, but relatively simple vehicle. Today's automobile uses less energy and creates less pollution, but it does so at the expense of much more refined design and much more sophisticated control. Indeed hundreds of today's products provide improved reliability and reduced energy consumption through sophisticated microprocessor control.

Governments in Canada are convinced that if the present standard of living is to be maintained, Canadian productivity must be improved through technological advances. They are providing incentives for technological research and development, particularly in areas such as electronics and computer-aided design and manufacturing; areas that are engineering- and technology- and labour-intensive. It is likely that the current efforts to promote 'high technology' in Canada will be at least partially successful and that the result will be improved employment prospects for engineers and technologists.

The energy-related mega-projects that were in the planning stages in the late 1970s were expected to cause shortages of engineering manpower during the 1980s and 1990s. With recent reductions in energy consumption, some have been deferred, but others are proceeding at a reduced scale or with a longer development period. As easily exploitable energy sources become depleted, it will certainly be necessary to develop the less-attractive ones, such as Arctic and coastal oil and gas fields and the oil sands. These developments will require new technologies and will no doubt provide a significant amount of engineering employment.

The above factors and others suggest that engineering employment prospects will be good during the next few decades, mainly because Canadian society is becoming increasingly technology dependent and technologically oriented.

ENGINEERING AND TECHNOLOGY ENROLMENTS

While engineering and technology programmes will be influenced by the anticipated decline in numbers of high-school graduates, it is anticipated that they will be moderated by other factors.

In the first place, the number of women entering engineering and engineering technology programmes has traditionally been very small, but in recent years it has begun to grow rapidly. For example, the number of female engineering students in Canada increased from practically none in the early 1960s to about 460 (more than 7 per cent of the total student population) by 1982/83. Groups of female engineers and university admissions officials are actively encouraging women to pursue engineering careers. Hence, it is anticipated that during the next decade the proportion of women in Canada's engineering programmes will exceed 20 per cent, the current proportion in the United States. A similar trend is anticipated in engineering technology programmes.

Engineering and engineering technology enrolments have tended to be more strongly influenced by employment prospects than have enrolments in other disciplines. Thus if, as has been argued, engineering employment prospects remain strong, engineering and technology enrolments should not experience the significant reductions that are anticipated in other disciplines. Canada's current rate of output of engineering graduates (25 per 100,000 of population) is low compared to that of some other industrialized countries such as Japan (56), the Federal Republic of

Germany (37), Hungary (51) and Poland (31), although it equals that of the United States.

By the same token, Canada's stock of engineering technologists is relatively small. While the annual number of engineering technology graduates currently equals the number of engineering graduates, there are at present more than twice as many engineers as technologists. The Society of Engineering Technologists of British Columbia has suggested that ideally, the engineering team should include from three to eight technologists for each engineer.

In summary, it is very likely that university and community college enrolments will decline by as much as 25 per cent during the remainder of this century, as a result of recent sharp declines in the birth rate and in immigration. However, with increasing emphasis on technology in Canadian society, engineering employment opportunities will probably remain good and this, combined with larger numbers of female students, will help to maintain engineering and engineering technology enrolments approximately at present levels.

THE ENGINEERING AND TECHNOLOGY EDUCATIONAL INSTITUTIONS

During the late 1970s and the early 1980s, Canada's post-secondary educational institutions had to cope with rapidly increasing enrolments while receiving only modest resource increases. Unfortunately, in the face of tight budgets, the universities at least allocated a reducing proportion of their resources to engineering. For example, between 1970 and 1979 the number of engineering degrees granted increased by 35 per cent, while the number of engineering academic staff increased by only 13 per cent. The corresponding increases for all disciplines were 29 per cent and 35 per cent respectively. The Dean of one of Canada's largest engineering faculties estimated that in the early 1980s, his faculty's budget was decreasing by 2 per cent per year in real terms.

For much of the period, the engineering units have managed to cope through increased staff productivity—larger classes, heavier teaching loads and heavier research supervision loads. However, by 1983, most engineering Deans agreed that further restraint would significantly degrade the quality of engineering education. Indeed, similar concerns have been expressed by the Canadian Accreditation Board, which has noted an increase in the number of short-term accreditations in recent years.

Of particular concern is the present state of obsolescence in Canada's engineering faculties. The Natural Sciences and Engineering Research Council, in its 1981 report on a survey of university equipment needs, estimated that during the following five years, funding of $115,000 per engineering staff member would be needed to restore research equipment to an adequate state. With the rapid advances in technology, particularly in the areas of computing, micro-electronics and computer-aided design and manufacturing, both engineering and technology institutions face momentous problems in trying to provide adequate modern instructional equipment.

In view of the projected rapid ageing of the population, it is almost certain that provincial governments will be channelling financial support away from the educational institutions and into services for the elderly. Even if the educational institutions direct increasing proportions of their resources into engineering, it is unlikely that those resources will be adequate to maintain the present standard of education. Hence, support will be necessary from other sources. The Natural Sciences and Engineering Council has substantially increased the value of its equipment grants to universities in recent years and that trend will likely continue. Perhaps more significantly, industrial firms have shown a tendency recently to be more supportive of educational institutions through donations of equipment, release time for their staff to permit them to teach part-time or supervise graduate students and to participate in industry/university co-operative research ventures. As Canadian industry continues to become less 'branch plant' oriented, its emphasis on research and development increases. Thus, it is anticipated that, increasingly, Canadian industry will view the engineering educational institutions as sources of research and development talent and, increasingly, industry will co-operate with and support the institutions.

Apart from their funding problems, the major problem facing Canada's engineering faculties today is that of recruiting qualified academic staff. Between 1958 and 1981, the percentage of engineering academics in Canada holding doctorates increased from 20.8 to 81.9. Unfortunately, a large majority of engineering professors completed doctoral research work on purely scientific subjects and on problems not necessarily encountered in engineering practice. Furthermore, few engineering academics have spent substantial parts of their careers in industry—the typical 25-year career pattern is 19 years at university, 5 years in industry and/or government and 1 year in other environments. Perhaps more ominous though, is the ageing of the engineering professoriat and the meagre supply of Ph.D. students in the system.

In 1982, 75 per cent of Canada's engineering professors were 40 years of age and older. The rate of attrition during the preceding five years had been sixty professors per year. On the other hand, the number of Canadians and permanent residents in Canadian engineering Ph.D. programmes declined throughout the late 1970s and by 1981 it was estimated that only 80 to 90 Ph.D. degrees were being awarded annually to that group. It was estimated further that about 40 Ph.D.s in engineering were being awarded to Canadians in other countries. Thus, at most perhaps 120 new Canadian Ph.D.s are available to replenish the stock of engineering professors and to meet the increasing needs of

Canadian industry for research and development personnel. While the situation looks somewhat bleak at present, there is perhaps some room for optimism. Following five years of decline, there was an increase of 20 per cent in the number of engineering doctoral students in 1981 and 1982. With the improved postgraduate scholarship programmes of the Natural Sciences and Engineering Research Council and the evident increase in industry/university co-operation, the trend is expected to continue.

The academic staffing problems of the engineering faculties are not shared by the engineering technology institutions. There, stronger emphasis is placed on industrial experience and the Ph.D. is normally not a requirement. Consequently, while the securing of qualified staff was sometimes difficult during the expansionary period of the 1960s, the problem at present is the securing of staff positions, not the filling of them. That situation is not expected to change within the next decade.

In summary, in the face of an ageing population, governments will almost certainly devote a decreasing proportion of their budgets to post-secondary education in the remainder of this century. In the face of, at most, modest increases in student numbers, it is highly unlikely that any new engineering faculties will be established (except for those in the University of Victoria and Simon Fraser University in British Columbia, both of which are in the process of being formed). Nor are the existing faculties likely to expand significantly. Similarly, further significant expansion of the engineering technology educational system is unlikely during the next two decades.

CURRICULA

One of the anomalies in Canadian engineering education in recent decades is that while engineering educators and employers seem to agree that undergraduate education should be basic and general (the student should be taught how to learn rather than how to do a specific job), the number of accredited programme titles tripled between 1970 and 1983 and the degree of commonality of programmes has decreased in most engineering faculties. On the other hand 80 per cent of all engineering degrees awarded in Canada in the 1970s were in chemical, civil, electrical and mechanical engineering. With today's rapid technological advances, there is strong pressure to provide more programmes, more specialized programmes and longer ones. Most engineering faculties today seem to deal with those pressures by maintaining four-year undergraduate programmes which include perhaps one half-year of technical elective choice, and leaving further specialization to graduate programmes. In addition, the trend toward more programme titles appears to have reversed in the United States and it is anticipated that this will happen in Canada. Furthermore, in order to prepare engineers to operate in fields in which knowledge is doubling every few years, it will be necessary for programmes to concentrate on the basics of mathematics, basic science and engineering science, with general approaches to design and synthesis.

With regard to the various major disciplines, electrical engineering is the one that has experienced the most dramatic changes within the past thirty years. During the past five years, electrical engineering student numbers have increased spectacularly, mainly at the expense of civil engineering. In addition, in half a dozen Canadian universities, computer engineering programmes have been 'spun off' from electrical. While electrical engineering will probably remain the most popular among the engineering programmes for some time, there is a danger that computer engineering will become too popular and very soon will be overstocked. On the other hand, computer applications will be stressed increasingly in all engineering programmes and in engineering offices.

While civil engineering enrolments declined sharply in the early 1980s, the field is likely to regain popularity during the rest of this century. In the first place, most of our multi-billion-dollar infrastructure of buildings, parking facilities, bridges, streets, highways, railways, terminals, sewer and water systems, treatment facilities and the like has been put in place during the past sixty years. Many of its elements will soon be in need of replacement, repair or upgrading. The work will entail a large civil engineering component. So too will the construction of air and water purification and waste treatment facilities that will be demanded by an increasingly environmentally conscious society. Finally, the inevitable energy-related projects of the future will provide employment for civils as well as others.

Mechanical engineering will probably remain almost as popular as electrical, and industrial engineering will probably increase in importance in response to the national drive for improved industrial productivity. The latter emphasis will no doubt also give impetus to mechanical, electrical, electronic and industrial technology programmes.

Summary

Engineering education in Canada began with a two-and-a-half-month civil engineering course at King's College, Fredericton, New Brunswick, in February 1854. By 1900 there were six engineering faculties in existence, none west of Toronto. The development of engineering faculties was gradual during the first half of the twentieth century, although by 1950 there were eleven engineering faculties between Halifax and Vancouver. The rapid industrialization of the country following the Second World War led to a doubling of the number of engineering faculties in the 1960s and early 1970s.

Engineering technology educational facilities were practically non-existent until 1960, when the Federal Government passed the Technical and Vocational Training and Assistance Act. Under that Act, the Federal Government agreed to reimburse the provincial governments 75 per cent of approved costs for the construction and equipping of technical training facilities.

In the late 1970s, the members of the post-war baby-boom generation (the sons and daughters of returning Second World War veterans) raised enrolments in the engineering and technology institutions to unprecedented levels. However, with the passage of that 'wave' through the educational system, enrolments are expected to remain constant or perhaps to subside during at least the next two decades.

Canada's birth rate has been declining sharply since 1960 and its immigration rate has been low since the early 1970s. As a consequence, the population is ageing; the median age increased from 26.3 to 29.5 between 1961 and 1981. It is expected to reach 37 by the year 2001. Cognizant of the population trends, governments began, in the 1980s, to redirect resources from education to services for the elderly. The trend is likely to continue through the decade. While the engineering and technology educational institutions have thus far been able to maintain high quality programmes through increased staff productivity, signs of impending deterioration are becoming evident. Equipment and facilities are badly in need of upgrading and a shortage of engineering staff with advanced degrees and industrial experience may well occur within five or ten years.

On the other hand, Federal Government programmes to upgrade university equipment and facilities and to encourage graduate study appear to be making an impact. In addition, a national will to improve Canadian productivity through improved technology is adding impetus to engineering research and development. A new spirit of co-operation between industries and educational institutions is developing and it will benefit both parties.

Canada's engineering and technology education system is complemented by a well-developed system of professional associations and technical societies. The result is a consistently high quality of engineering and technology education and practice. Though resource limitations threaten the education quality at present, judging by its past performance, the Canadian engineering community will cope successfully with this latest challenge.

Acknowledgements

The assistance of the Deans of Canada's engineering faculties and of representatives of the following organizations is gratefully acknowledged: Canadian Council of Professional Engineers; Engineering Institute of Canada; Association of Professional Engineers of Manitoba; Society of

Certified Engineering Technicians and Technologists of Manitoba; Ontario Association of Certified Engineering Technicians and Technologists; Canadian Council of Engineering Technicians and Technologists; University of Manitoba; Red River Community College; Natural Sciences and Engineering Research Council.

Bibliography

A Five-Year Plan for the Programs of The Natural Sciences and Engineering Research Council. Ottawa, Natural Sciences and Engineering Research Council, 1979.

A Statement on Engineering Education by the Canadian Council of Professional Engineers. May 1982.

Annual Report. Canadian Council of Professional Engineers. 1982.

BAIRD, A. Foster. The History of Engineering at the University of New Brunswick. *The University of New Brunswick Memorial Volume*. Fredericton, University of New Brunswick, 1950.

BARTON, T. H. *Engineering Employment—Past, Present and Future*. The PEGG, Association of Professional Engineers, Geologists and Geophysicists of Alberta. October 1983.

BROKENSHIRE, E. G. *A Survey of Community College Systems in Canada*. 1980. (Task Force on Education, Background Report No. 8.).

BRONET, F. Bridging the Gaps; Women in Engineering. *Engineering Journal*, Volume 66, No. 2, March 1983.

CALDWELL, R. T. The Historical Development of Red River Community College. *Manitoba Technologist*, Volume 8, No. 3, June 1978.

The Canadian Council of Professional Engineers. *Newsbrief*, Volume 23, No. 6 (E), June 1981.

Canadian Council of Professional Engineers 1983 Synopsis of Provincial and Territorial Professional Engineers' Acts, By-Laws and Procedures. 1 January 1983.

COLLINS, P. *Notes on the Centenary of the Faculty of Engineering of McGill University*.

Commonwealth Universities Yearbook 1982, Volume 2.

EGGLESON, Wilfrid. *National Research in Canada*. Clarke, Irwin & Company Ltd, 1978.

EMMERSON, George S. *Engineering Education: a Social History*. Newton Abbott (UK)/New York, David & Charles/Crane Russak, 1983.

The *Engineering Journal* in Retrospect. *Engineering Journal,* May 1943.

Engineering Manpower News. Canadian Council of Professional Engineers, Ottawa, Ontario.

FOOT, D. K. *Canada's Population Outlook; Demographic Futures and Economic Challenges*. Published by the Institute for Economic Policy, Suite 409, 350 Sparks Street, Ottawa, K1P 7S8.

The Future of Education and Careers in Technology. Society of Engineering Technologists of British Columbia, January 1981.

HARRIS, Robin S.; MONTAGNES, Ian. *Cold Iron and Lady Godiva, Engineering Education at Toronto 1920–1972*. University of Toronto Press, 1973.

HOWARD, Cyril H. *A Survey of the Independent Learning Practised and Planned in Non-University Post-Secondary Institutions in Canada*. March 1978.

JONES, L. E. (ed.). *The Next Hundred Years*. Faculty of Applied Science, University of Toronto, 1973.

KIRBY, C. C. *A History of the Dominion Council*. Canadian Council of Professional Engineers, 1946.

KONRAD, Abram G. *Clientele and Community. The Student in the Canadian Community College*. Association of Canada Community Colleges, 1974.

LEAL, H. Allan; CORRY, J. Alex; DUPRÉ, J. Stefan. *The Report of the Professional Organizations Committee*. Ministry of the Attorney General, Government of Ontario.

MASCOLO, Dominique. *Engineering Education in Canada, Some Facts and Figures*. A preliminary draft of a report for the Science Council of Canada, April 1983.

MORGAN, L. M. *A Summary of Accredited and Approved Programs*. Edmonton, The Alberta Society of Engineering Technologists.

MORRIS, G. A. The Canadian Accreditation Board. *Proceedings, Third Canadian Conference on Engineering Education, Saskatoon, Saskatchewan, May 1982*.

SEAFORD, J. Two Categories: Two Classifications. *Manitoba Technologist*, Volume 13, No. 3, June 1983.

SLEMON, J. R. *Technology Transfer: Government/Industry/University*. Paper presented at the International Electrical, Electronics Conference Exposition, Toronto, 28 September 1983.

The Story of The Engineering Institute of Canada. *Engineering Journal*, June 1937.

Twenty-Fifth Anniversary Year Book 1945. Association of Professional Engineers of the Province of Manitoba.

The University of Manitoba, Faculty of Engineering, 75th Anniversary Volume, July 1982.

YOUNG, C. R. *Early Engineering Education in Toronto 1851–1919*. University of Toronto Press, 1958.

Case-study for Japan

Lawrence P. Grayson,
National Institute of Education,
Washington, DC 20208, USA

Contents

Technological education in Japan

Introduction

In recent years, Japan has emerged as the most vibrant economic force in the world, building an economy to a level and at a rate unmatched in modern economic history. Since the end of the Second World War, Japan's Gross National Product has grown almost sixtyfold, to a per-capita level second only to the United States. Today, Japan accounts for about 10 per cent of the world's economic activity, although occupying only 0.3 per cent of the world's land surface and supporting about 2.6 per cent of the world's population.

In 1950, Japan's GNP per capita (a measure of how much each person produces and, ultimately, of how well he lives) was 6.5 per cent that of the United States; by 1984, it was 75 per cent. In 1984, Japan had a $30.3 billion surplus in world trade. During the fifteen years from 1968 to 1983, the average hourly pay of Japanese workers in manufacturing grew more than ninefold, which is a direct reflection of the nation's 108 per cent increase in industrial productivity.

Japan is in a strong position to continue its progress. Its GNP has increased from $10 billion in 1950 to $1.1 trillion today. It has an extensive base of modern, capital-intensive industries and is building new industrial capacity at a more rapid rate than most other nations. Further, it is well positioned to compete for a larger share of the world market, with an extensive structure for exports, a large trade surplus and a strong currency.

This position of pre-eminence was not due to a set of fortuitous circumstances, but rather was the result of a carefully planned effort at social change and adaptation to a world dominated by technology. There are many reasons why the Japanese have made such tremendous economic progress. They have a clear economic policy that has guided government actions. They have the highest rate of personal savings and capital formation among the major industrialized countries. Since equity financing in Japan is primarily through banks and only minimally through public stock offerings, corporate management can focus on long-term growth rather than short-term profits (of the world's thirty largest banks, eleven are Japanese). The nation spends little on defence. Government and industry work together closely, and trade practices protect the home market. Further, a strong work ethic and company loyalty have been fostered by the almost paternalistic attitudes of industry.

Another key factor, however, is Japan's strong commitment to education, particularly engineering education. Since the latter half of the nineteenth century, when modernization began, education has been prominent in the policies of successive governments. Japan has stressed the expansion of scientific and technical education at all levels, both to provide the engineers and technicians needed by industry for growth and technical development, and to produce the technically literate population required to facilitate the transfer and adoption of technology on which the nation's industrialization has depended. As a result, Japan has been able to assimilate with ease new ideas and techniques from other countries and adapt them to fit Japanese needs.

Educational development and economic growth have been linked reciprocally in Japan. The increased quantity and quality of education have provided the manpower necessary for industrial development. In turn, economic growth created the necessity for, and enabled the country to afford, the great expansion in the number of upper secondary school and college graduates.

In the past thirty years Japan has supported, and several times revised, education as part of its plans for economic development. During this period industry demands have significantly affected educational policies. This influence is especially important since Japan's success rests on the continued development of technology-based industries, which require people with an advanced level of knowledge and access to technical information. That education could become an instrument of economic policy stems from the close relationship between business and government in Japan.

Development of educational policy

The period after the end of the Second World War, from 1945 until mid-1951, was devoted first to demilitarization, then to reform and democratization of Japanese society, including education, and then to the beginnings of economic rehabilitation. Education was restructured after American models. The number of educational institutions, particularly colleges and universities, rose significantly, and advanced education was made more widely available.

After Japan gained its independence in May 1951, the process of adapting American ways to Japanese needs and culture began. A major concern addressed immediately was the development of education as a means for promoting economic growth. One month after Independance, the National Diet passed the Industrial Education Promotion Law, which recognized that 'industrial education is the basis of the development of the industry and economy of our country'.[1] Private industry stressed the same theme as Nikkeiren (the Federation of Employer Organizations), an influential group representing major economic organizations in Japan, and called for revisions in education to make it more directly relevant to industry's needs.

In a series of reports issued between 1952 and 1959, Nikkeiren recommended the promotion of scientific and vocational education in elementary and middle schools, the improvement of technical training for working youths, the creation of five-year technical colleges that would combine vocational high schools with two years of college, and more training in college-level engineering and science. Technical colleges were seen as essential for training the middle-level personnel required by industry. The organization's concern went beyond domestic development and recognized that continued industrial and economic growth would depend on competing in the international market-place:

> If Japan does not develop systematic training of engineers and experts in the midst of its epoch-making economic growth in order to attain further progress in industrial technology, our industrial technology will lag day by day far behind the international level. This will result in our failure in competition with other nations.[2]

The government responded to the demands of industry. In 1957, the Economic Planning Agency, established to co-ordinate policies among government departments with regard to long-range economic planning, estimated that 27,500 science and technology graduates would be needed annually by 1962. Later in 1957, the Ministry of Education issued a five-year plan for expanding the number of science and technology graduates by 8,000 and announced major curriculum revisions for elementary and secondary schools.

The coupling of economic and education policy continued, as the government adopted in 1960 the National Income Doubling Plan, which set the goal for the decade. The Economic Council of the Economic Planning Agency, which drafted the plan, underscored the importance of education when it stated that 'economic competition among nations is a technical competition, and technical competition has become an educational competition'.[3] The Economic Council recommended immediate action to educate a large number of scientists and engineers so that the economic plan would not be handicapped by a lack of human resources.

The Ministry of Education, acting in harmony with the economic goals, developed a plan in 1961 to raise the number of places in science and technology faculties from the then 28,000 to 44,000 within seven years. As part of this enlargement, nineteen new five-year technical colleges, comprising three years of upper secondary school and two years of college, were established in 1962.

As the educational system expanded, planners realized that education should fill a broad range of purposes. In 1963, the Economic Council proposed a diversified, but meritocratic system of education. Its report, which made recommendations for all levels of education, including education in industry, again called for expanding programmes in science and technology. This strongly affected educational policy planners.

The ten-year goal of doubling national income was achieved in seven years, and Japan experienced unprecedented prosperity. Although several readjusted economic plans were developed in the 1960s, the educational plans remained the same.

One reason why education policy and its implementation could be so responsive to science and technology policy is that the Ministry of Education was organized in the 1960s to receive advice from the scientific community. Reporting to the Minister of Education in 1966 were a Higher Education and Science Bureau, a Science Encouragement Committee, and a Science and Education and Vocation Education Council. In addition, the ministry had direct responsibility for the budgets of the national universities, colleges and junior colleges, their attached research institutes and a National Training Institute for Engineering Teachers, giving it a great deal of influence with those institutions.

In 1970 the Economic Council published its new economic and social development plan, which again broadened the purposes of education. First priority remained the improvement of science and technical education. To cope with the increased internationalization of industrial activities, the report suggested teaching the knowledge, skills and other necessary qualities needed to develop international co-operation,[4] a theme that continues to receive increased attention.

In June 1984, the University Chartering Planning Sub-committee of the University Chartering Council published a report that continued the planning for higher education begun in 1976. As the sub-committee developed plans for higher education up to the year 1992, one of its basic considerations was the continued economic development of the nation. As Japan has reached a very high level of development in science and technology, a goal proposed for universities is to educate 'excellent and creative researchers and other personnel who can explore new fields of study'.[5]

EMPHASIS ON ENGINEERING

Since the end of the Second World War, Japan's economic strategy has targetted certain industries for growth and assistance to make them globally competitive, with each newly identified industry being more technologically advanced (and potentially adding higher value to the economy) than previous ones. The Japanese have become major exporters, first of textiles, then of steel, automobiles and consumer electronics, and now of semiconductors. They are the world's industrial leader in robotics and optical electronics, and are making gains in computers, telecommunications and biotechnology, industries that will be important for the remainder of this century for any nation aspiring to high economic growth.

In semiconductor memory chips, Japan's progress has been rapid. In 1970, when the 1 K RAM chip (a random-access memory chip) that can store up to 1,024 bits of information) was standard, Japan had virtually none of the world market. In 1974, when the 4 K RAM was the state-of-the-art, Japanese manufacturers had about 5 per cent of the world market. With the introduction of the 16 K RAM, the Japanese gained about 40 per cent of the world market by 1978, and with the 64 K RAM, Japan's share rose to 70 per cent by the end of 1981.[6] That is significant, not only because of the amount of the sales revenues, which were estimated to be $1 billion in 1984 and are projected to reach $2 billion in 1986, but because the 256 K and the future 1 M chips are essential to successful competition in the computer, telecommunications and other high-technology industries, as well as in a wide variety of consumer products for personal, entertainment and household uses.

To provide the necessary technical manpower to achieve their nation's goals, economic and education policy-makers in Japan have supported increases in the number of engineering schools, engineering enrolment and faculty members. In the 30 years from 1955 to 1984, the number of bachelor's degrees in engineering awarded in Japan rose from 9,613 to 70,486.

Engineering education, however, is only one aspect of Japan's commitment to developing human resources and preparing its people and industries for a high-technology-based society. The nation has a strong elementary and secondary education system whose graduates are highly knowledgeable and supportive of science and technology; a higher education system that produces a large pool of engineering and technical manpower; and extensive in-company education and training to make employees effective in meeting both short-term and long-term goals of the corporations. Industry's general satisfaction with the educational system is attested to by Hosai Hyuga, chairman of Sumitomo Metal Industries, Ltd, Japan's sixth largest exporter of technology in 1980, who stated: 'The high level and nature of the Japanese education system makes it very easy to turn a high school graduate into an auto assembly line worker or a college graduate into an electronics engineer'.[7]

This structure is supported by a culture and society that places a high premium on achievement in education, an emphasis that permeates Japanese society. It is reflected at all levels of students from the preschooler to the employed adult, in the industrial policies for manpower development and the cultural concern for academic achievement, in the emotional support of the student by his family with its *kyoikumama* or 'education mama', and by the hundreds of thousands of *ronin* students who, having failed to gain admittance into a preferred school will continue to study and reapply the following year or even for several years. Educational achievement determines one's opportunities through life and is a central factor in national development.

As a result of the nation's intense desire for education at all levels of society and an education system that maintains high standards, Japan has produced what probably is the best-educated population in the world. Although global assessments are difficult to make, in several international comparisons in mathematics and science conducted since 1964, no country has surpassed the Japanese in overall mastery of subject-matter. Not only is the average achievement of Japanese students very high, but their range of performance is concentrated at a higher level than almost any other nation. As a result, Japan has a well-educated labour force receptive to learning specialized skills or being retrained at the workplace.

GENERAL EDUCATION

Japanese education is composed of six years of elementary school, three years of lower-secondary and three of upper-secondary school, and four years of college. Although only nine years are required, over 94 per cent of the students continue to upper-secondary school, and almost all of them (95 per cent) complete that level. Classes in Japan meet five-and-a-half days a week for thirty-five weeks or more a year. The school day, which runs from 8.30 am to 3 pm, is divided into six subject-matter periods for five days and four periods on Saturday.[8] Although 210 school days a year are required, most schools are open 240–250 days a year.

Further, many concepts and skills are taught at greater levels of difficulty and at earlier ages than in other countries. National school standards in Japan require that over 25 per cent of the class time in the elementary and lower-secondary schools be devoted to mathematics and science—subjects introduced in the first grade and continued throughout the nine years of compulsory education. Elementary-school students are taught such mathematical concepts as correspondence of geometric figures, and probability and statistics. In upper-secondary school, the curriculum is divided into two streams. Students in the academic course of study, which is completed by about 30 per cent, take three years of mathematics, including at least the elements of probability and statistics, vectors, differentiation and integration, and the functions and programming of computers.[9] Students in the non-academic and vocational stream may also complete three years of mathematics. They study the same subjects, although their courses have an applied orientation. Further, at more than 100 upper-secondary schools, students interested in pursuing careers in science and mathematics may elect an intensive curriculum that includes eighteen to twenty credits of mathematics.

A similar emphasis prevails in the sciences. In upper-secondary school, many students complete three natural sciences (physics, chemistry, biology, or earth science). Courses in Japan have drawn heavily on American biology, chemistry and physics curricula, such as those of the Physical Science Study Committee (PSSC). Materials initially were imported, seminars held with American specialists, textbooks prepared and laboratory equipment adapted to Japanese requirements. Newsletters, workshops and published papers promote the teaching approaches. Japanese scientists and science teachers also have developed new types of laboratory equipment, often substantially funded by the Ministry of Education, and at local initiative established science education centres in each of the forty-six Prefectures (or districts) throughout the country, where thousands of teachers have received in-service training in mathematics and science. Foreign-language teaching begins in seventh grade, with about 10 per cent of class time devoted to it. Although students have a choice of languages, almost all elect to take English.

By the end of upper-secondary school, most of the college-bound students have studied six years of a foreign language. All Japanese students, including those who become craftsmen or production workers or pursue other vocational trades, are well educated in mathematics and science. The average factory worker in Japan is more likely than his counterparts in other countries to be capable of discussing and implementing quality-control techniques based on sampling procedures, describing in analytic terms a production problem on the shop floor, and being retrained easily for more skilled jobs that may be created by automation.

The education system in Japan is deeply affected by an extreme reliance on examinations to determine advancement. The first-level college entrance examination administered nationally every year is the primary determinant for admission to college. Excellent performance is necessary if a student is to be eligible to take the entrance examinations of the more prestigious universities. This series of examinations tests scholastic achievement in the Japanese language, two areas of social studies, mathematics, two sciences, and a foreign language (in 1980, 99.7 per cent of the students chose English). Since in the Japanese system graduation from college is virtually guaranteed after acceptance, and many large businesses limit their hiring to graduates of certain prestigious institutions, the college entrance examination in effect determines future opportunities and success.

Teaching, as a result, is geared to preparing students for the college entrance examination. As most upper-secondary schools also require subject-matter achievement tests for admission, lower-secondary schools have become examination-oriented, a pattern that continues downward throughout the system. Many parents send their children to extra-study schools known as *juku*, where attendance ranges from only a few per cent of students in early elementary school to 40 to 50 per cent of seniors in upper-secondary school. The desire to attend the better upper-secondary schools is so great that a new class of *ronin* has been established, as students who fail to gain admission to preferred upper-secondary schools devote another year to preparation rather than accept a school of lesser quality.[10]

Although the emphasis on examinations channels teaching toward the most able students and leaves little opportunity for flexibility in subject-matter, it has produced for Japan a highly effective education system, especially in mathematics and science. In 1970, Japanese youth in both the 10- and 14-year-old age-groups scored first among nineteen countries in each of a series of international science tests in biology, earth science and chemistry, and physics.[11] In the First International Project for the Evaluation of Educational Achievement, conducted in 1964, which compared the abilities of students from twelve industrialized nations, the Japanese 13-year-olds ranked first in mathematics, with 76 per cent scoring in the upper half of the scale. Japanese students also were the most positive in their liking of mathematics.[12] Since the comparison is of scores earned two decades ago, these students are now in their prime years in the labour force. In the second international assessment of mathematics conducted in 1984, Japanese seventh-grade students scored first in each of the five categories tested, competing against eighth-grade students from nineteen other nations.[13] For countries competing in the international market with products and services that depend on advances in modern technology, these results are significant.

TECHNICAL AND JUNIOR COLLEGES

There are two types of mid-level institutions in Japan that educate technicians—technical colleges and junior colleges. Although both began shortly after the Second World War, they were started for different reasons and have come to serve different functions.

Shortly after the economic recovery began, Japanese industry recognized the need for mid-level technicians highly trained in the practical application of science and technology. In 1962, the Ministry of Education created nineteen five-year technical colleges that combined the three years of upper-secondary schooling with two years of college. Among their purposes was to provide basic knowledge and techniques of design, production and construction, as well as the ability to plan and execute work in areas of management, testing, research and surveying.[14] Although the number of technical colleges rapidly increased to fifty-four by 1965 (to provide uniform distribution throughout Japan), their growth since then has been modest. In 1984, 57 technical colleges enrolled 47,527 students in engineering and graduated 7,869. Although there are twenty recognized major disciplines at technical colleges, two-thirds of the students major in mechanical, electrical or civil engineering, with the remainder divided among aircraft, graphical and production engineering, industrial design, information engineering, etc.

In some ways, technical-college courses resemble those taught at engineering colleges. In mechanical engineering, for example, there are courses on heat transfer, fluid dynamics, fluid machinery, automatic control, instrumentation and measurement, and dynamics of machines, as well as the properties of plastics and other non-metallic materials. The difference, however, is that the courses stress practical applications, rather than theory. Faculty members are recruited from among technicians in industry, as well as among persons with teaching experience in college and upper-secondary school. Technical colleges were developed as a terminal level of schooling, but about 10 per cent of the graduates continue their education at four-year colleges.

In response to the strong demand for technicians with practical knowledge, plans are being developed to improve the course content of existing departments in colleges of technology and to expand them. Further, two universities of technology and science were founded in 1976 at Toyohashi and Nagaoka to provide graduates of technical colleges with opportunities for continued education. About 20 per cent of the students enter the four-year institutes after graduation from upper-secondary school, while the remaining 80 per cent enter at the third-year level after completing a technical college. As these new universities combine practical training with education, every fourth-year student is assigned to a position with an affiliated company for training at a job site.

In contrast to technical colleges, which were established in response to industrial demands, junior colleges were begun during the occupation as part of the democratization process initiated by the allies. Junior colleges, which are patterned after the two-year American model, are attented after the completion of upper-secondary school. The Japanese education establishment was slow to accept the junior colleges, and they were not recognized under education law as permanent institutions until 1964. In 1984, junior colleges graduated 6,516 people with engineering training, almost as many as the technical colleges. Although curricula are similar, junior-college graduates are not highly esteemed in society. Technical-college graduates receive a degree, while junior-college graduates are awarded a certificate. Since many large companies will not hire persons with junior-college training, they tend to become technicians in medium- and small-sized companies that serve as subcontractors to large companies.[15] The low prestige of junior colleges may reflect the status of women in Japanese society. Of the 169,000 junior-college graduates in 1984, 90 per cent were women, almost 30 per cent of whom majored in home economics. Junior-college education in Japan is considered a suitable preparation for marriage and homemaking, but not for industrial employment.[16] In industry women have had few opportunities for permanent employment or advancement, a situation that may be gradually starting to change.

The structure of higher education

Higher education in Japan is built on a foundation of meritocracy and élitism, which produces a high-quality education for the able, but also structural rigidity and inbreeding. Although schools for advanced education date back several hundreds of years, universities in the current sense are relatively recent. Tokyo University, the first 'true' university in Japan, was created by the government in 1877 to meet national aims—principally to educate high-level government officials and a technically élite group responsible for absorbing technology from the United States and Europe. To achieve these ends, the university selected the most able people (through examinations) and educated them well. Since the preparation of government administrators was the primary function of the university, the Faculty of Law had the highest status, although engineering, economics and agriculture were also important because of their relationship to technical developments.

To meet a growing demand for education, Kyoto University was established in 1897. Universities grew slowly in the next fifty years, as the government's promotion of modernization focused on developing a highly educated élite. General literacy was left to the elementary and secondary schools. At the outbreak of the Second World War in 1940, there were only forty-five universities in Japan, more than half of which were private. Tokyo and Kyoto Universities had the responsibility of educating political leaders, the other national universities focusing on training the technical élite, while the private colleges met the general demand for higher education. The basic structure and élitism that resulted from this division remains today.

Beginning under the allied occupation, in order to democratize education and educate a large segment of the population, the number of private colleges and universities in Japan has expanded greatly, while the number of state-supported national universities has grown much more slowly. By 1982, there were 455 universities, of which 95 were national universities, 326 were private and 34 were local universities established by the prefectures and municipalities.[17]

Although under the law all universities are equivalent and grant the same degrees, in practice they are far from equal. The rapid expansion required the upgrading of many lower-level schools, including some of the upper-secondary schools, into colleges and universities. Large numbers of new faculty members had to be recruited from a small base of qualified people. The older, established universities had excellent resources and graduates in influential positions in government and industry, and had well-earned reputations for quality. Further, the government's policy, stated in 1963, was to diversify education according to student abilities, aptitudes and the needs of society.[18] This policy helped to maintain and reinforce the stratification and élitism among schools.

Today, a dual system of higher education exists in Japan. One group of institutions consists of low-tuition national universities and a very few high-quality private universities. These educate about 20 per cent of the college population and are usually characterized by strong research activities and graduate courses. The second group is made up of a large number of high-tuition private institutions and local universities that educate the rest of the college population. Their main function is undergraduate instruction. Differences between national and private universities are striking. As shown in Table 1, national universities surpass private universities by almost every quantitative indicator.[19] Very few of the best of the private institutions can match the resources of even the poorest of the national universities. The exceptions are Waseda, Doshisha and Keio, and perhaps a few of the older, private universities that enjoy a higher status and provide a better education than many of the newly created national universities.[20]

Although 33 per cent of Japanese upper-secondary school graduates pursue further education, competition for admission—especially for the more prestigious universities—is keen. The Tokyo Institute of Technology, for example, bases its admission decisions not only on the applicant's performance on the national college entrance examination, but on its own 6½-hour series of written examinations in mathematics, physics,

TABLE 1. Comparison of national and private universities, 1982

	National	Private
Number of institutions	95	326
Students	425 141	1 339 877
Full-time teachers	49 850	51 622
Part-time teachers	21 575	45 382
Ratio of students to full-time teachers	8.5	26.0
Floor space of school buildings (in thousands of square metres)	14 025	15 694
Size of school sites (in thousands of square metres)	1 334 932	125 193
1980 expenditure (in billion Yen)	1 062	1 328
Expenditure per student (in million Yen)	2 497	991
R&D expenditures in engineering (in billion Yen)	183.1	134.6
Baccalaureate graduates undertaking further education	15.2 %	2.3 %
Tuition	(Low)	(Moderate)

chemistry and a foreign language. Nationwide, in 1983, only 19.1 per cent of the applicants to engineering schools were admitted. Of those, 74 per cent graduated from upper-secondary school that same year, while the rest were *ronin*, 22 per cent of whom graduated the previous year, and 4 per cent two or more years earlier.[21]

Higher education, however, is less demanding than the entrance examinations would suggest. After the intense studying to pass the entrance examinations, students are granted a respite from continued pressure. Students rarely are failed or academically challenged to the same degree as in high school. Faculties are highly autonomous and maintain a great deal of power in most universities. They protect their budgets and academic domains tenaciously. Curricula are rigidly prescribed, new fields of study are difficult to start, and cross-disciplinary studies are almost impossible to have approved. There is little opportunity for innovation in universities.

Engineering curricula

Colleges and universities in Japan continue to play an important role in economic development. The Japanese graduate a substantial number of people with concentrations in engineering, management or administration backgrounds. Today, 20 per cent of the baccalaureate-level students in all universities, and 26 per cent in the prestigious national universities, graduate with engineering degrees, and another 3 per cent graduate in natural sciences. Among engineering students, 29 per cent are educated at national universities, 69 per cent at private universities, and less than 2 per cent at local universities. Students in Japan are distributed among engineering disciplines, including architecture, as shown in Table 2.

In curriculum matters, requirements for a bachelor's degree in engineering in Japan resemble those of schools in the USA. This similarity is not surprising, since the major educational reforms occurred

TABLE 2. Engineering degrees by field (percentages), 1982

Field	Bachelor's	Master's	Doctoral
Electrical and electronics	26.4	22.7	21.2
Civil and architecture	24.4	15.0	13.0
Mechanical	21.4	16.9	12.9
Chemical	11.5	19.3	22.9
Other	24.2	26.1	30.0
Total number of degrees per annum	73 593	7 363	621

after the Second World War. The allied occupational forces brought in groups of American advisors, including one engineering education group. As the Japanese attempted to follow the lead of the United States in technological development, many Japanese students have studied in the USA since the 1950s and brought back to Japan the best of what they found.

In addition to courses in science, the humanities and engineering, Japanese universities require twelve or more credits in the study of one or two foreign languages. This training, following six years of foreign-language study in secondary school, gives Japanese professionals a strong reading ability in a second language, and allows them to read the literature of at least the English-speaking countries. This facility is capitalized on at the universities. Of the 276,000 volumes in the University of Tokyo's engineering library, for example, only 111,000 are in Japanese or Chinese. Waseda University has a library of over 1 million volumes, of which slightly more than 30 per cent are in Western languages.

This language ability fits well with the nation's approach to development. Since the Japanese have historically pursued the course of primarily adapting Western technologies, rather than relying on original research and innovation, they must know what other countries are doing. Japanese ministries, large industries and trading companies expend a huge effort in information gathering and analysis. Vogel has described the scope and importance to the Japanese of these activities:

> If any single factor explains Japanese success, it is the group-directed quest for knowledge. In virtually every important organization and community where people share a common interest, from the national government to individual private firms, from cities to villages, devoted leaders worry about the future of their organizations, and to these leaders, nothing is more important than the information and knowledge that the organizations might one day need. . . . It is not always clear why knowledge is needed, but groups store up available information nonetheless on the chance that someday it might be useful.[22]

Graduate schools were not developed before the Second World War. They existed, nominally, but there was no prescribed programme and no fixed period of residency for candidates. Students pursued their own research interest under a senior professor, while they awaited appointment to the faculty. After the war, under American influence, Japanese universities began offering master's degrees in the mid-1950s and doctorates in the early 1960s. The growth of graduate programmes has been slow, however, as industry has not shown a keen interest in hiring people with doctorates.

In 1963, Japan graduated 693 engineers with master's degrees. By 1982 that number had risen to 7,363. At the doctoral level, the number of engineering Ph.D.s granted in Japan increased from 83 to 621 over the same nineteen-year period. Employment opportunities for Ph.D.s in Japan, however, are limited.[23]

University faculties in the older and more prestigious universities operate under an expanded *koza* system, in which collections of chairs have been grouped for administrative, teaching and research purposes. Faculties are relatively small but numerous. A university frequently will have several faculties with similar names that teach the same or similar subject-matter. Each chair or *koza* is an administrative unit which traditionally consists of one professor, one assistant professor, and one or two assistants, all selected by the professor but approved by the faculty conference (i.e. departmental assembly). A chaired professor holds absolute authority over those under him. They must support his interests totally; his betterment means the betterment of the group. This structure is reinforced by the strong group orientation of Japanese society. Young faculty members defer to their elders in virtually all matters. Chair-holding professors closely and restrictively control the granting of the Ph.D. As competition among chairs for funds, space, new assistants and the most able students is intense, and since the success of the group depends on how well it works together, isolation and inbreeding result. In order to reduce the power and autonomy of the chairs, some of the newer universities are instituting the departmental system in which policy and general administrative matters are decided by groups of faculty members acting in concert, rather than by the most senior individuals.

The lifetime employment system means that there effectively is immediate tenure upon being hired as an instructor. There also is virtually no movement of faculty members to universities considered less prestigious than their current one. As one can rarely expect to be employed by a university higher in prestige than one's alma mater[24] there is great inbreeding at the universities of highest esteem. In 1975, 90 per cent of the professors at Tokyo University and 86 per cent at Kyoto were alumni of those institutions.[25] Even at newer national universities, 40 per cent of the assistant professors were alumni of the same institution in which they taught. The less prestigious institutions must rely on younger faculty members or on older ones who have retired or have not been accepted into prestigious institutions. Graduates of Tokyo and Kyoto Universities have come to occupy 38 per cent of all university positions in Japan.[26]

Further, until September 1982 all foreign nationals were prohibited by law from holding a regular faculty position in a national or public university. Even since the passage of the new law, very few universities have hired foreign nationals as faculty members. These factors are serious for both the teaching and research functions of the university, for productive scholarly activity requires the free flow of ideas and information. Japan recognizes this shortcoming and is planning to recruit foreign students actively at the graduate level on a greatly expanded basis.[27]

Training in industry

Industry plays an important part in the overall educational process; lifelong education and on-the-job training are integral parts of the policies of Japanese industry. Under the lifetime employment system of large companies and government agencies, upgrading the qualification and skills of employees is essential for the future vitality of the organization.

In the early twentieth century, as Japan was beginning its industrialization, there was a severe shortage of qualified personnel and a high turnover in most types of labour. To retain a skilled work-force, large companies evolved a system of lifetime employment. Salaries were determined by seniority rather than performance, many benefits including health services and recreational facilities were provided, and many activities of high motivational value were introduced to help employees identify their future with the company's. As lifetime employment evolved, Japanese companies—as a strategy for the future—became willing to invest substantially in developing their employees.

Today, most large companies, including Nippon Electric, Toyota, Matsushita, Hitachi, Fujitsu, Sony and others, run company schools and educational institutions for all employees—blue-collar, clerical, engineers and other professionals, and managers. The education that is provided may take the form of general education, such as learning how to program a computer, or company-specific training, such as the procedures used for manufacturing video displays.

New employees in Japan typically are hired in April each year as they graduate from college. They are not employed for a particular project or a given job, but are hired to be employees of the company as a whole. They are viewed as a group or class, and with regard to salary and titles they progress at about the same rate for fifteen years or so. For good performance, they are given new assignments, provided with job security and good working conditions, and eventually will be promoted. At an early stage, those recognized to have potential for high-level management are given assignments, but not necessarily the title or salary, to develop their potential. Eventually, they rise to high-level positions. The lengthy period of equality not only binds the group together, but allows ample time for employees to develop and demonstrate their abilities. By the time they reach their late thirties or early forties, the class begins to spread, as some of the members rise into middle and senior management. When they are about 55, most members of the group will retire and only those who have been identified for top executive levels of management remain.

Typically, newly hired engineers will spend from a few weeks to six months or more in classroom and job-site training under the supervision and guidance of more experienced engineers. Only then will they begin work, under the continued supervision of senior staff. During this period and through their careers, they can engage in in-plant seminars, professional conferences and on-the-job training. When they are in junior and senior management positions, they most probably will attend company-conducted schools, either for management or technical training. During this period they also might be chosen to participate in a survey tour abroad to learn first-hand about certain technologies or management techniques in other countries, or for a company-sponsored scholarship. Many large companies and government ministries typically send some of their brightest young employees abroad for advanced education.

Under the lifetime employment system, it is economically sound for large organizations to provide some young employees with extended periods of training in areas related to their future responsibilities. Typically, after they are employed for a few years and have learnt the needs of their organization, a few of those with potential for technical or managerial leadership are sent abroad for several years to study at the best foreign universities, with their salaries and expenses paid by the organization. Many study in fields related to technology, finance and, particularly, business. This training allows them to develop a knowledge of foreign business affairs and make contacts that will help the future leaders throughout their careers.

Japanese companies have not been inclined to hire people with doctoral degrees. They prefer to hire baccalaureate or master's graduates and train them in-house to meet company goals. This practice has several advantages for the company. First, it is believed that baccalaureate and master's graduates are more malleable than persons with a doctoral degree, as the latter frequently wish to remain in their areas of speciality. The company orientation and early training programmes are expected to have more effect on younger graduates in helping them to identify with the company, rather than maintain a strong allegiance to their disciplines. Secondly, this training can be tailored to meet the company's specific needs, particularly in technical areas. Finally, the person with training provided by the company lacks the advanced credentials that might make him more employable in an open market (as limited as it is in Japan), and thus less committed to the company. As a result the Ministry of Education reported that in 1982 of recent graduates in engineering 1 in 46 baccalaureate-level and 1 in 58 Master's level graduates were unemployed, but 1 in 7 Ph.D.-holders had not found employment.[28]

No universal pattern or policy for employee development or continuing education is followed by all companies. Each Japanese company is free to develop its own priorities and policies consistent with its own objectives and knowledge of good business practices. Large companies operate their own education and training activities; some are quite extensive. Medium and small companies, however, must rely on business organizations or associations for educational programmes.

Matsushita is one company that invests heavily in employee development. Virtually every one of the over 120,000 employees has had substantial company-provided training. The firm makes no particular effort to recruit from the élite universities, in contrast to most Japanese companies, as it believes in hiring young people who can be easily trained and starting them at the bottom. Every professional, whether engineer, accountant or salesman, begins by spending six months in training status, which includes selling or working directly in a retail outlet, in a factory performing routine tasks on an assembly line, and attending lectures by senior executives on 'the Matsushita way'. After the initial training, employees are assigned to divisions of the company, where they continue to take courses from its Research and Training Institute. Mentoring and job rotation to gain a broader overview of the company figure in the employee-development programme. When professional or managerial employees are to be promoted, they will study at the institute for almost a year to upgrade technical and managerial skills, and for reinforcement in the company spirit. Every employee is taught the corporate philosophy, organization, procedures and management system. These shared understandings and common belief that their most important activity is to meet customer needs with marketable products at minimum cost have been extremely important to Matsushita as it has grown and diversified its product line.

Matsushita is not unique, however, in the amount and type of training it provides. The Nippon Electric Company, for example, provides continuing education for professional employees at its Institute for Technical Education, while Hitachi has established the Hitachi Institute of Technology where more than 5,000 company engineers were trained in the last ten years. The Toyota Motor Company, the Tokyo Electric Power Company, many major banks and hotels, as well as large companies in the service area are among the many companies that conduct extensive in-house training.

The inevitability of change

The remainder of this decade will be significant for the Japanese. International response to the nation's level of exports, the maturing and stabilization of its economy, and the rapid ageing of its population and work-force are particularly notable trends that may require significant changes within Japan, if the nation is to maintain its economic vitality. These changes will very likely occur in its business practices, its social structure, its view toward research and its education system.

Japan's success in exporting goods is directly affecting other nations' industries. As a result, many countries are beginning to place limitations or restrictions on the import of Japanese products. Japan may have to moderate its exports, if further trade barriers are not to be imposed. As Japan relies heavily on exports, this will directly affect its economy.

In addition, the economic growth of Japan is beginning to stabilize. For the fifteen-year period from 1955 to 1970, the average annual increase of the nation's real GNP was 10 per cent. For the decade of the 1970s, however, the average annual growth rate was slightly less than 5 per cent and for the period 1980–84 it was 4.2 per cent.[29] Japanese economists predict that the nation's economy will continue to grow at about this same rate for the remainder of the decade. Although this is greater than the predicted world economic growth rate of slightly less than 4 per cent, it is less than during the country's high growth period. A more stable economy implies fewer job openings due to growth and less need for new, young workers in Japanese industries.

Further, the population is ageing at a rate so rapid that it is unprecedented among modern nations. Japanese longevity has increased from an average life expectancy of 50 years in the mid-1930s to 70 years today. To exacerbate matters, since the 'baby boom' after the Second World War, fewer children are being born to bear the social and economic costs of an ageing population. In 1950, only 5 per cent of the population was 65 years of age or older. Today, about 9 per cent of the population is over 65 years of age, and that group will increase to 14 per cent by 2000 and about 20 per cent by 2020. This rapid ageing of the population has very significant implications for Japan. There will be increased social costs. Larger numbers of retirees imply larger outlays for retirement payments by industry, greater medical costs, more leisure and housing facilities for the aged, as well as other social benefits. In Japan currently, fifteen working people support each senior citizen; by 2015 it will be just three.[30]

The work-force also is ageing rapidly. Under the traditional Japanese wage structure, in which salary is correlated with age, this implies higher average wages and greater labour costs, raising the prices of products and making them less cost-competitive in international markets. This system was very beneficial to industry in the 1950s, when one-quarter of all workers were under the age of 25. The rapid expansion of industry and the influx of new young workers kept the percentage of employees under the age of 25 to 21.6 of the labour force as late as 1970. Slower growth, however, reduced the proportion of young workers to 13 per cent of the labour force by 1978, and the average age of employees continues to rise.[31] Matsushita Electrical Industrial Co., for example, estimates that if there is no change in its pay system, the ageing work-force will cause its payroll to rise 30 per cent in the next ten years.[32]

FUTURE DIRECTIONS

How must Japan change to account for these trends? It is a country with few natural resources, that currently imports a majority of its energy, and that relies on other nations for an increasing share of its food. Since Japan cannot depend for its future economic growth on agriculture or on the manufacture of goods whose production is energy-intensive, labour-intensive or consumes large quantities of natural resources, it must rely on the development of advanced technologies. One official of the Ministry of International Trade and Industry (MITI) has said that in the future '*All* industries have to become smarter both in the way they make things and in the amount of knowledge the products themselves contain'.[33]

The strength of Japanese industry has been in adopting fundamental knowledge from other countries, improving on it, and in designing, manufacturing and marketing the resulting products. In order to obtain the basic knowledge, Japan has relied on licensing agreements with other countries. In almost all of the past 30 years and in each year since 1975, Japan has increased over the previous year the amount of technology imported,[34] with over 50 per cent of the technology introduced into Japan coming from the United States. As the state of technology changed, the types of Japanese imports changed. In 1975, for example, 9.6 per cent of the agreements were for computer hardware and software; by 1980, 22.2 per cent were for the same items.

Japan's strategy of identifying and transforming existing fundamental knowledge into products for the market-place has been eminently successful. In early 1970, sales of products derived directly from imported technology accounted for more than 30 per cent of the industrial sales in Japan, and since that time imported technology has contributed increasingly to the nation's production of goods for export.[35] Japan, however, can no longer rely primarily on the United States or Europe for basic knowledge. As Japan competes in fields such as semiconductors, computers, telecommunications and biotechnology, where the state-of-the-art is changing rapidly and in which organizations that do the basic research and development have a competitive advantage, its strategies will have to change. For the first time, Japan is in the position of having to advance the state-of-knowledge, do advanced research and create its own technologies.

The Industrial Structure Council, a policy advisory body to the Ministry of International Trade and Industry, foresees that Japan must develop a creative, more knowledge-intensive industrial structure. This must be based on the capability for original technological developments, and produce higher-value-added products by focusing on software and knowledge intensification.[36] A Nikko Research Institute report on potential industrial growth areas for the 1980s stated the need as follows: 'Japan has to develop its own technologies by means of the superior brains of its people—its only resource—in order to ensure its economic survival'.[37]

In short, the nation must rely on its 'brainpower'. Yet, in a survey conducted by the Science and Technology Agency in April 1984, most Japanese enterprises felt that they were lagging behind their European and American competitors in their ability to develop technology. One of the main reasons given was 'lack of talented people'.[38] The intellectual challenge that faces Japan, and which Japan poses to other countries, is how to educate and utilize people who can create and produce new, advanced technologies that will give Japan a competitive education in worldwide economic competition.

NEED FOR CREATIVITY

Creativity is essential for technological leadership. The Japanese recognize this, and are intent on fostering innovative thinking. Takuma Yamamoto, President of Fujitsu, summarized the views of many of his countrymen with his statement: 'The creativity of the Japanese people will be called into question from the latter half of the 1980s to the 1990s. The whole nation must work like one possessed to meet this great challenge'.[39]

In the past, however, there have been very few examples of Japanese creativity in original technical development. The areas in which Japan has excelled, such as advances in semiconductors, VLSI (Very Large Scale Integrated circuits), fibre optics, and carbon fibres have been due more to continued improvements rather than technological breakthroughs. Japan to date has stressed pragmatic thinking in its R&D, seeking product quality and reliability, rather than originality.

This approach is consistent with Japan's social orientation. Group identity and harmony are extremely important factors in Japanese society, influencing not only business practices and organization, but family relations, religious practices, social organizations, and interpersonal relations.[40] Creative people, however, are often mavericks who set themselves apart from the group because of strong commitments to their beliefs and ideas. For the Japanese, fostering creativity could create social conflicts. Computer specialist Yasuo Kato has stated that the Japanese 'are not so creative because the creative mind is peculiar, and we Japanese don't like anything peculiar. We believe that everyone should be the same'.[41]

Social uniformity, however, need not preclude change. Japan has shown itself on several occasions to be highly adaptive and has adopted practices that have been in conflict with the existing social system. When the Japanese were awakened to the advances of the West by Commodore Perry's visit in 1853, they rapidly changed from a feudal, agricultural society that was closed to outsiders to one with a manufacturing and industrial base having expansionist ambitions. After the nation's defeat in the Second World War, the Japanese readily adopted American ideas and institutions, modifying them to fit the nation's needs, and a period of rapid economic growth began. There is no inherent reason why the Japanese cannot again modify their social system, if the need is strong enough.

CHANGES IN EDUCATION

Perhaps nothing is more fundamental to Japan's technological development and economic growth in this century than education. The Japanese have created an extremely fine education system that has produced a highly literate population, very homogeneous in its basic knowledge, with a good understanding of mathematics, science and other subjects that are central to modern technology. Yet, there are significant problems in the education system, particularly at the university level.

Fostering creativity, for example, is intertwined with questions of education. If Japan is to develop personnel capable of achieving technological breakthroughs, it must stress individuality, research, originality and risk-taking. These are not the characteristics, however, that are promoted in the present education system. As Takehiko Matsuda, President of the Tokyo Institute of Technology has remarked:

> Students merely learn to answer a question rather than how to ask a question . . . The Japanese students of engineering, it seems to me, are generally very well trained in solution techniques for differential equations, for instance, but do not possess the problem-oriented mentality and skills to formulate the equation.[42]

The significance of the single university entrance examination creates great pressure on the student to learn the types of information that are included in the examination. This has the effect of distorting education. Much of the schooling in senior high school is devoted to preparing the student for the university entrance examination, rather than to learning in the broader sense. Former Minister of Education Michio Nagai states: 'It is not an exaggeration to say that education designed to develop men who love learning and think for themselves has already been abandoned'.[43]

Although a certain amount of local autonomy is allowed at the elementary and secondary levels, the high standards for student achievement, published curriculum guidelines, and control of the certification of the textbooks by the Ministry of Education promote a condition of uniformity in which children in the same grade, throughout the country, learn the same *katakana* characters and the same way to solve equations at about the same time. The Japanese classroom, at the university as well as at the pre-university levels, is characterized by rote learning, copying of lecture notes, and an almost absolute reliance on testing and test results. In Japanese schools, says Tokyo University professor Takemitsu Hemmi, 'students don't have to be able to discuss. They just say, "Yes, I understand" '.[44] This produces students who learn great amounts of information, and typically far out-perform students from other countries when tested in traditional ways.

In spite of the rapid assimilation of large amounts of information, or perhaps because of the system that produces it, the Japanese have not been noted for creativity and originality in their thinking. Masanori Moritani, a senior researcher at Nomura Research Institute finds that

> what Japan hungers for today is not a uniform crop of gifted students but people with extraordinary talents, heterodox people, human resources with the potential for achieving great things . . . we [the Japanese] are entering an age when we must strive to search out and pinpoint the extraordinary and unorthodox among us . . . there is a real danger that these individualists will be numbered among the dropouts in today's education system. The question of how Japanese industry can best discover and train men and women of this disposition will be one of the greatest issues facing it in the future, on which will hang the success or failure of growth and development for years to come.[45]

Although recruitment of university graduates for government and business positions usually is by examination, there is a very strong correlation between one's performance on these examinations and the university one attends. This probably is due more to the fact that the largest firms limit their recruiting to the most prestigious universities and to the native ability of the students, rather than the quality of their university studies. Those who score highest on employment examinations are most likely to be the ones who, four years earlier, scored highest on the university entrance examinations. The fact that the needs of Japanese industry are met as well as they are, in spite of the problems of higher education, is due in large measure to the high quality of pre-university education and to the pre-service training programmes conducted by industry and government.

Recognizing that education will continue to be an essential element in Japan's strategy for development, MITI has proposed that for the future well-being of the nation, Japan must establish an 'age of vitalized human potential'.[46] Although this is a generally accepted goal, it is not clear what this means or how it will be achieved. There already is a trend in Japan toward higher levels of education for both men and women. Since 1970, persons over 20 years of age who have completed higher education have increased from 11.5 to about 16 per cent of the population. By 2000, one in four Japanese is expected to have completed post-secondary education.[47]

Although there is a great increase in the number of students receiving baccalaureate degrees, relatively few people complete graduate studies. Advanced graduate education currently is undertaken primarily by persons preparing for academic careers. Although Japan graduates more engineers at the baccalaureate level than does the USA (even though Japan has only one-half the population), Japan is graduating about half the number of master's and one-third the number of doctoral candidates in engineering as the United States. In spite of the difference in numbers, the USA cannot meet the demands of its industries and academic institutions for persons with doctorates, while there is not enough demand in Japan for persons educated to this level. The situation is similar in the sciences. The total number of people receiving master's and doctoral degrees in physics and chemistry in Japan is only about one-quarter that of the USA. As a nation desiring to base its economy on knowledge industries and striving to achieve world leadership in high technology, Japan will require significantly more people with graduate education, particularly Ph.D.s in engineering and science.

RESEARCH ACTIVITIES

In order to achieve its goal to become a knowledge-intensive nation, Japan will have to foster environments that can lead to sustained success in extending the state of knowledge in many scientific and technological fields. The striking feature of Japanese research, however, is its direct relevance to the needs of industry and commerce. This is true of research supported by the government, as well as business. Industry's concept of research and development is exemplified by Matsushita, where a primary purpose of its twenty-three production research laboratories is 'to analyse competing products and figure out how to do better'.[48]

University-based research activities in Japan are extensive. In 1984, 1.0 trillion Yen, which was 15.8 per cent of the total R&D funds expended nationally, was spent at universities. Industry accounted for 4.6 trillion Yen, which was 70.1 per cent of the R&D funds expended, and an additional 14.1 per cent was expended at research institutes, some of which are attached to universities.[49]

Research activities in Japan are not well integrated with the teaching functions of the university. At many university-affiliated research institutes separate faculties are appointed exclusively for the research functions. They attain a status within the university that is independent of, but equal to the teaching faculties.[50] As research is an important element of graduate education, the teaching and research functions of the university must be more closely integrated if graduate education in Japan is to be increased.

In addition, much of the government-supported research is not well co-ordinated among organizations. Most of the funds expended for science and technology come from the Ministry of Education. Of the 1.44 trillion Yen allocated by the government for science and technology in the 1983 budget, 49 per cent was provided by the Ministry of Education, Science and Culture, 27.1 per cent by the Science and Technology Agency, 11.8 per cent by the Ministry of International Trade and Industry, 2.7 per cent by the Defence Agency and 9.4 per cent by other sources.[51] Unfortunately, there are strong rivalries among the ministries, and many of the ministries that support R&D do not co-operate with each other on many matters. They do not even relate well to the Science and

TABLE 3. Comparison of university expenditure (in billions of yen) by source for engineering research in Japan, 1983

	Expenditure	Source (%)	
		Government	Non-government
National universities	561	97.5	2.5
Private universities	418	13.8	86.2

Technology Agency, which was established to plan and co-ordinate the government's efforts in these areas. Space science research, for example, is conducted at the University of Tokyo under support of the Ministry of Education. Space application R&D is carried out by the National Space Development Agency, under the direction of the Science and Technology Agency. The two organizations have separate launch facilities and do not engage in any interchange of scientific personnel.[52]

The pattern of research support is quite different between national and private universities in Japan. As noted in Table 3,[53] almost all of the funds at national universities are provided by the government, while 86 per cent of the engineering research funds expended at private universities are provided by industry and other non-government sources. Although it would appear that private universities in Japan have stronger links with industry, it must be recalled that the national government has a strong planning and co-ordination role with regard to industry, and the national universities can serve a very valuable function in a three-way partnership as the government attempts to develop future economic opportunities for the country. In a recent White Paper, the Science and Technology Agency advocated interchange and co-operation among industries, universities and government research institutes.[54]

Japan recognizes the need to stimulate its creativity, and already is supporting basic research in areas such as Very Large Scale Integrated circuits (VLSI), fibre optics, communications systems, voice-recognition systems, and other advanced electronic technologies, as well as in energy conservation and pollution control for which there are strong needs in Japan. Furthermore, the Industrial Structure Council of MITI has set numerous objectives for the 1980s, including both the further development of knowledge-intensive and innovative technologies, such as microcomputers, optical communications, VLSI, and laser beam devices, and the creation of next-generation technologies in the life sciences, energy, and data processing. Particular emphasis will be placed on creating new materials, developing large-scale systems for alternative energy sources, and creating technologies for social systems, such as for personal and community activities.[55]

In order to stimulate creativity on a more national scale, the government and industry are jointly supporting two major activities. The High-Speed Computer Project is an attempt to develop computers through the use of new materials and designs to operate at speeds 100 to 1,000 times the speed of today's fastest machines. The Fifth-Generation Computer Systems Project is a major attempt to do world-class research and develop new types of computers, utilizing techniques of artificial intelligence and the concept of expert systems, which are capable of processing information symbolically. The latter is not only an attempt to place Japan in a significant technological position as the world's leader in computers, but is a major experiment in fostering basic research and creativity.

CHANGES IN HUMAN RESOURCE DEVELOPMENT

Company-provided education and training play an important role in Japanese industry. Employees of Japanese firms are not hired to work at a particular job or within a particular unit. Rather, they are hired as employees of the company, who will work in a variety of jobs in various parts of the organization throughout their careers. With lifetime employment granted to many employees of large firms, vacancies at all but the lowest levels are filled by internal promotions. In order to be effective, this internalized labour market requires that workers not only obtain broad experience within the firm, but receive appropriate training at various stages in their careers. This training is important both for motivating employees and for developing the needed human capital within the firm.

Significant increases in higher levels of education, the ageing work force, and the internationalization of Japanese business, however, raise serious problems for industrial education and training, and for the types of employment that will be available for Japan's already highly qualified work-force. Greater numbers of college graduates are leading to a surplus of these people, and to underemployment; people now are having to accept jobs that previously were filled by others with less education. In 1978, almost 55 per cent of the university graduates in industry, in the group 40–44 years of age, had advanced to the level of division or department manager; by 1988, only 30 per cent of this group will reach these levels.[56] Thus, many employees will have to settle for lower-level positions. Even in the most successful age-group, only 30 per cent of the university graduates are expected to attain management positions in 1988. The remaining 70 per cent most likely will retire from non-management jobs.

If longevity no longer guarantees promotion, new forms of worker evaluation, such as that based on performance or achievement, and education and training will have to be devised. Merit-promotion systems have been used at companies such as Matsushita and Canon for many years, and now are being adopted by more companies. Today, the notion of using examinations as the primary means of assessing ability is gaining currency. The examination system is seen as an objective method of detecting workers who may be lacking in a particular skill, which could then be provided through appropriate training and job assignments.

Although the leading Japanese firms have developed rather elaborate training programmes that emphasize total and comprehensive personal development, most formal education programmes are provided for new employees who are recent school graduates. Twice as many Japanese firms offer education programmes for the recent graduate than for either present employees or new employees who have prior experience with other firms.[57] The majority of the jobs in Japanese industry were designed, in the past, for young workers who constituted the majority of the workers. Now tasks must be remodelled to fit the older worker. As this is not simply a question of finding a job, but of productively utilizing the worker, it must be viewed within the larger context of career development and training.

As Japan's economy has grown, both its exports and its overseas investments have increased rapidly. This massive expansion of trade and investment represents a significant internationalization of the activities of Japanese industries. Direct overseas investment is different to trade, for it involves conducting a business under local rules and regulations. It is essential that any firm engaging in international activities develop a comprehensive and well structured plan for the training of its personnel. Training must be provided not only for those employees who will work or manage the business in another country, but for the many employees in the parent company who will deal with the overseas operations. In large companies, training programmes, which may extend over several years, may include overseas job rotation and instruction on the customs, culture, history, legal system, and political and social systems in the country, as well as language training.

Fujitsu has an interesting approach to its development programme for businessmen to serve in international activities: 20 per cent of the management training these employees receive is devoted to cultural subjects, including music, art, drama, literature, history, religion, and poetry, as well as international politics and international economic activities. Since the early training and experience of the employee has been devoted to improving his technical competence, the company now stresses those subjects which will give him the personal skills and breadth of view needed to deal both formally and informally in a wide variety of situations with people from other countries.

The future role of women

One of the most significant changes in Japanese society occurring from the emphasis on technology and education will be the assumption of new roles by women. Although women comprise one-third of the labour force in Japan, they have occupied primarily clerical, assembly or temporary positions. Their expendability has been a significant factor in allowing large corporations to maintain the lifetime employment system for men. The degree of education and the aspirations of women, however, are changing. College-educated women are projected to increase from 6.4 per cent of the female population in 1970 to 20 per cent in 2000.[58] Higher levels of education coupled with a labour shortage in certain career specialities are creating new opportunities for women. The Industrial Structure Council, for example, estimates that today Japan requires 795,000 software engineers. Yet, there are only about 100,000 of these specialists in the country, and high-level male engineers are reluctant to remain in this field because generalists in industry have better promotion opportunities. Industrial recruiters are now beginning to go to women's colleges to hire software engineers.

Venture capital firms, although few in number at present, also are creating new options for women. Entrepreneurial firms have great difficulty recruiting highly qualified men, who prefer to enter large firms and rarely leave them once they are hired. Women, regardless of their qualifications and ambitions, have had almost no opportunities in large firms to rise to management positions, as these jobs have been reserved exclusively for men. Women, however, are being given management roles in new companies.[59] With significant numbers of women entering the professional work-force, there will be pressure to have more women advance to management-level positions in small and large firms. This will exacerbate a projected shortage of management-level positions for university graduates, and will create new relationships between men and women. The effects of these relationships will extend well beyond the firm.

The changes that now are beginning in many areas of Japanese society will accelerate as Japan continues to reach a more stabilized, but advanced level of economic development. Prime Minister Yasuhiro Nakasone spoke of the extent of those changes when he said: 'We [the Japanese] must formulate a society for which there is no precedent in any other country'.[60] The future of Japan will depend on how well the nation can effect the changes it plans and deal with the desired and undesired consequences of those changes.

Summary and perspective

Education in Japan has been an instrument used to achieve economic ends. That has been true since industrialization began over 100 years ago, but it has been particularly used in this way since the 1950s. Government policies and private industrial goals have become intertwined, providing harmony among political, national development and private economic motives. This has enabled the state to initiate policies and take actions that directly enhance the development of industry and enlarge corporate profits. Industry in turn has provided the jobs, products and services required for rapid economic growth.

As the economy expanded, industry required, demanded and received, first, large increases in the number of engineers being educated, then the creation of new technological colleges to produce middle-level technical personnel and, more recently, changes in curricula to reflect a broader international awareness. While the size, direction and rapidity of the educational development are due to the economic and political imperatives, the education system that has evolved has been shaped by the culture of the country. This has produced an education system that is in accord with the Japanese temperament, and has facilitated economic growth by satisfying the needs of industry for entry-level manpower.

The education system has served Japan's industrial needs well for the last thirty years. Guidelines from the centralized Ministry of Education, Science and Culture, frequently developed in response to the nation's economic plans, and a strong reliance on the nationally administered college entrance examination as the primary factor in determining college admissions, have fostered uniform and high standards among the elementary and secondary schools of the country. Graduates of the secondary schools are well-grounded in science, mathematics and other subjects, and provide Japanese industry with a well-educated labour force receptive to further training at the work-place. Technical colleges and junior colleges graduate the technicians for large- and medium-sized companies, while four-year colleges and universities educate engineers and other professionals. Industry provides continuing education and training throughout the working life of an individual, both to increase his general intellectual growth and to make him more knowledgeable about the needs and procedures specific to the company. Education, in its entirety, has provided the intellectual base necessary for the rapid and large economic growth of the country.

Conditions in Japan, however, are changing. The period when the nation could rely almost exclusively on applying and improving on ideas developed in other countries has come to an end. Japan must now stress individual creativity and initiative as it bases its future industrial development on an intensified use of knowledge.

Japanese industry has become the leader in several fields of high technology, primarily by importing basic knowledge and ideas discovered in other countries and using them to develop and manufacture new or improved products. This approach may not continue as a fully successful strategy for the future, as the Japanese will be required to develop their own technologies. Continued economic growth may demand significant changes in Japan's present strategies for technological development and, in turn, to its education system.

The beginnings of change already are evident. There is a growing national consensus that Japan should alter its policies for science and technology from emphasizing adoption to promoting creativity and inventiveness. The government is declaring that Japan will become a technologically oriented nation, rather than a trade-oriented one. As it moves towards becoming a technology-based, knowledge-intensive nation, education will remain an essential element of Japan's development strategy. The Japanese will stress intellectual achievements over natural resources and physical labour, as they strive to maintain the vitality of their economy. The attitude of Japanese business towards education was well expressed by the president of Nippon Telegraph & Telephone Public Corporation (NTT) when the stated: 'High intelligence is the only source of competitiveness.'[61]

Notes

1. Tetsuyu Kobayashi, *Society, Schools and Progress in Japan,* p. 90, New York, Pergamon Press, 1976.
2. Nobuo K. Shimahara, *Adaptation and Education in Japan*, pp. 129–130, New York, Praeger, 1979; Kobayashi, op. cit., p. 91.
3. Shimahara, op. cit., p. 133.
4. *Basic Guidelines for the Reform of Education.* Report of the General Council of Education, Ministry of Education, Japan, 1972.
5. University Chartering Planning Sub-committee of the University Chartering Council, *The Systematic Planning and Administration of Higher Education in Japan After 1986*, p. 13, Ministry of Education, Science and Culture, 6 June 1984.
6. *The Effect of Government Targeting on World Semiconductor Competition, a Case History of Japanese Industrial Strategy and Its Costs for America*, p. 39, Cupertino, Calif., Semiconductor Industry Association, 1983.
7. 'Japanese Technology, The Cutting Edge', *Fortune*, 23 August 1982, p. 24.
8. *Education in Japan, a Graphic Presentation*, p. 59, Tokyo, Ministry of Education, Science and Culture, 1982.
9. *Course of Study for Upper Secondary Schools in Japan*, pp. 8–9, Tokyo, Ministry of Education, Science and Culture, 1983.
10. William K. Cummings, *Education and Equality in Japan*, pp. 213–15, Princeton, N.J., Princeton University Press, 1980.
11. L. C. Comber and John P. Keeves, *Science Education in Nineteen Countires*, p. 159, New York, John Wiley & Sons, 1973.
12. Torsten Husen (ed.), *International Achievement in Mathematics (a Comparison of Twelve Countries)*, pp. 187 et seq., New York, John Wiley & Sons, 1967.
13. F. Joe Crosswhite et al., *Second International Mathematics Study, Summary Report for the United States*, Figs. 23–27, January 1985.

14. *Technical and Technological Education in Japan*, p. 78, Japanese National Commission for Unesco, December 1972.
15. Ibid., p. 62.
16. Ronald S. Anderson, *Education in Japan: A Century of Modern Development*, pp. 216–18, United States Department of Health, Education and Welfare, 1975.
17. *Statistical Abstract of Education, Science and Culture, 1983 Edition*, p. 66, Tokyo, Ministry of Education, Science and Culture, 1983.
18. Kobayashi, op. cit., p. 101.
19. *Statistical Abstract of Education, Science and Culture, 1983 Edition*, op. cit., pp. 66, 69, 76, 100–1, 106–7, 166–7, 170–1; *Indicators of Science and Technology (1984)*, pp. 84–5, Tokyo, Science and Technology Agency, 1985.
20. *Higher Education and the Student Problem in Japan*, p. 36, University of Tokyo Press, 1972.
21. *Japan Statistical Yearbook, 1984*, Table 19–17, p. 660, Statistics Bureau, Management and Coordination Agency, 1984.
22. Ezra F. Vogel, *Japan as Number One: Lessons for America*, p. 27, Cambridge, Mass., Harvard University Press, 1979.
23. *Indicators of Science and Technology (1984)*, op. cit., pp.116–17.
24. William K. Cummings, Ikuo Amano and Kazuyuki Kitamura, *Changes in the Japanese University, a Comparative Perspective*, pp. 152–3, New York, Praeger Publishers, 1979.
25. Akira Arimoto, *The Academic Structure in Japan: Institutional Heirarchy and Academic Mobility*, p. 28, New Haven, Conn., Higher Education Research Group, Yale University, August 1978. (Working paper YHERG-27.)
26. Anderson, op. cit., p. 186.
27. University Chartering Planning Sub-committee . . ., op. cit.
28. *Statistical Abstract of Education, Science and Culture, 1983 Edition*, op. cit., pp. 96–7, 100–1.
29. *Economic Report of the President, February 1985*, p. 356, Washington, D.C., US Government Printing Office, 1985.
30. 'The Social Impact of a Graying Population', *Business Week*, 20 April 1981, p. 72.
31. Haruo Shimada, *The Japanese Employment System*, p. 13, Tokyo, Japanese Institute of Labour.
32. 'An Aging Work Force Strains Japan's Traditions', *Business Week*, 20 April 1981, p. 81; 'A Changing Work Force Poses Challenges', *Business Week*, 14 December 1981, pp. 116–17.
33. 'Japan's High-Tech Challenge', *Newsweek*, 9 August 1982, p. 48.
34. Terutoma Ozawa, *Japan's Technological Challenge to the West, 1950–1974: Motivation and Accomplishment*, pp. 19, 23, Cambridge, Mass., MIT Press, 1984; *Indicators of Science and Technology (1984)*, op. cit., pp. 132–3, 140, 144–5.
35. R. Z. Caves and M. Vekusa, *Industrial Organizations in Japan*, Chap. 7, Washington, D.C., Brookings Institution, 1977.
36. *The Vision of MITI Policies in the 1980s, Summary*, p. 21, Tokyo, Ministry of International Trade and Industry, 17 March 1980. (Report No. NR-226 (80–7).)

37. *Japanese Industries, New Technologies and Potential Growth Areas in the 1980s*, p. 1, Tokyo, Nikko Research Center Ltd, October 1980.
38. *Toward Creation of New Technology for the 21st Century* (White Paper on science and technology in 1983), p. 13, Tokyo, Science and Technology Agency, December 1984. Distributed by the Foreign Press Center, Japan, as report W–84–20, January 1985.
39. Masanori Moritani, *Japanese Technology: Getting the Best for the Least*, p. 184, The Simul Press (Japan), 1982.
40. Chie Nakane, *Japanese Society*, Berkeley, Calif., University of California Press, 1972.
41. 'Electronics Research: A Quest for Global Leadership', *Business Week*, 14 December 1981, p. 89.
42. Takehiko Matsuda, 'Recent Developments and Future Trends in Engineering Education in Japan', *Journal of Engineering Education in Southeast Asia*, Vol. 14, No. 1, June 1984, pp. 5–6.
43. 'Schooling for the Common Good', *Time*, 1 August 1983, p. 21. (Special issue on Japan.)
44. Ibid., p. 67.
45. Moritani, op. cit., p. 179.
46. *The Vision of MITI Policies in the 1980s, Summary*, op. cit., p. 27.
47. Long-Term Outlook Committee, Economic Council, Economic Planning Agency, *Japan in the Year 2000, Preparing Japan for an Age of Internationalization, the Aging Society and Maturity*, p. 7, Tokyo, Japan Times Ltd, January 1983.
48. Richard Tanner Pascale and Anthony G. Athos, *The Art of Japanese Management, Applications for American Managers*, p. 42, New York, Warner Books, Inc., 1981.
49. *Indicators of Science and Technology (1984)*, op. cit., pp. 48–9.
50. Nakayama Shigeru, 'The Role Played by Universities in Scientific and Technological Development in Japan', *Journal of World History*, Vol. 9, No. 2, 1965, p. 354.
51. 'Science and Technology R&D Expenditures', *Science & Technology in Japan*, Vol. 2, No. 6, April/June 1983, pp. 40–1.
52. Justin L. Bloom, 'Japanese Science and Technology, the View from the Other Side', *Speaking of Japan*, August 1981, p. 28.
53. *Indicators of Science and Technology (1984)*, op. cit., pp. 48–9.
54. *Towards Creation of New Technology for the 21st Century*, op. cit., p. 20.
55. *The Vision of MITI Policies in the 1980s, Summary*, op. cit., pp. 15–16, 34.
56. Tadashi Amaya, *Human Resource Development in Industry*, pp. 5, 7, Tokyo, Japan Institute of Labour, 1983.
57. Shimada, op. cit., p. 21.
58. Long-Term Outlook Committee . . ., op. cit., p. 21.
59. Edward Boyer, 'Start-Up Ventures Blossom in Japan', *Fortune*, 5 September 1983, p. 118.
60. 'All the Hazards and Threats of Success', *Time*, 1 August 1983, p. 20. (Special issue on Japan.)
61. 'Hisashi Shinto: Discovering America's Secret', *The Washington Post*, 10 April 1983, p. F1.

Case-study for Pakistan

Professor Mohammad Authaulah,
Dean of Engineering,
NWFP University of Engineering and Technology,
Peshawar, Pakistan

Contents

Engineering and technical education in Pakistan

Introduction

Pakistan gained its independence as a sovereign state on 24 August 1947. The country has a variety of physical features including plains, mountains and deserts and has an area of 804,000 square kilometres. It has a population of 83.6 million, of which about half are below the age of 20. Nearly three-quarters of the population live in villages. The country has four provinces, namely Punjab, Sind, North-West Frontier Province (NWFP) and Baluchistan. As regards population, Punjab is the largest with 58 per cent, Sind 22 per cent, NWFP 17 per cent and Baluchistan 3 per cent of the total population. English is a compulsory subject taught from the sixth class to degree level. The national language is Urdu, but there are many regional languages spoken in various parts of the country.

Immediately after Independence the country was faced with staggering problems in the management of education. The first task was to save the system from collapse and to make good damage that had been caused by the turmoil following the partition of India and Pakistan. The magnitude of this task could be seen from the fact that, for a total population of 32 million, there were 8,430 primary schools, 2,190 middle schools, 408 high schools, 40 colleges and 2 universities. This situation provides a bench-mark for measuring educational development since 1947.

Education is a sacred asset and a driving force of a nation. The plans of economic development can only be laid on a sure foundation of technological education. In this fast-changing world, the raising of the quality of life is a major problem. The development of science and technology through educational effort is a key factor in the necessary economic growth. It calls for a system of manpower and training to raise productivity wherein special emphasis is placed on engineering and technical and vocational education.

Science and engineering and technology are considered to be the effective tools for development. There is no economic sector which does no reflect the intensive need for application of science and engineering and technology. The scientific and engineering effort has varied with time and with the state of development. It also varies between developed and developing countries. At present, economic development in the developing world is more technology- or engineering-orientated. This is evident from the fact that the development is based largely on industrial application and development rather than on other sectors of the economy.

Pakistan inherits many British traditions and concepts. However, during the last decade there has been a considerable change. Unlike Britain, where even now much, though by no means all, of the study of engineering occurs in universities historically founded upon the twin cultures of liberal arts and pure science, engineering education in Pakistan has now been separated from general universities and takes place in separate universities of engineering and technology. This move followed the bad experience of very slow development of constituent engineering colleges in general universities, where they were considered as having no greater status or importance than a science or arts department.

The exploitation, processing and putting to use of the natural resources of the country are engineering functions and their development is intimately linked with the development of industry. The type of engineer needed in Pakistan is one with vision and leadership, determination to grapple with local problems, and having a broad-based training in engineering. An engineer may be called upon to perform functions in any engineering project: research, development, design, construction and management. It is towards these ends that the engineering education process is guided, planned and operated. A clear distinction is made between the function of an engineer and that of a technician, and separate education programmes for each have been developed.

Industry needs not only engineers but associated groups of higher technicians and technicians (sometimes called engineering technologists in North America). The progress of technical education and training since 1947 has resulted from various steps taken as a result of the national policy on education. This policy produced a series of development programmes which can be divided into the following four distinct phases: Phase One (1947–59): Policy formation and implementation; Phase Two (1960–69): Expansion and consolidation; Phase Three (1970–77): Innovation and experiment; Phase Four (1978 onwards): Ideological reorientation.

Technical education was also considered as an integral component of overall education and so education has also undergone qualitative and quantitative changes during these phases.

Educational structure

The need for technical manpower differs according to level. At the lower level of technical skills, there is a great need for skilled workers. Pakistan needs more skilled workers than technicians. Similarly, the number of engineers and technologists will be comparatively less than the number of technicians. A generally accepted ratio of 1 : 5 : 100 between engineers, technicians and skilled workers determines the aims of manpower development. In order to match the education system with technical-manpower requirements the framework of education and training has evolved in order to provide a large number of technically qualified personnel at lower levels of technical skills, and few personnel at higher levels.

Education is a shared responsibility between Federal and provincial governments, but administered mainly by the latter. Technical education is controlled by provincial governments through Directorates of Technical Education. The system of vocational and technical education is integrated by liaison committees of directors of technical education. Engineering colleges and universities are controlled and funded by the Federal Government through its University Grants Commission.

ELEMENTARY EDUCATION

Primary education of five years' duration forms the basic educational structure. At this stage, activities conducive to overall growth and the development of the child form the content of the syllabi. The secondary stage, which is also of five years' duration, is divided into two phases. The first phase involves classes six to eight and is a continuation of primary education though with more purposeful ideas. At the end of class eight, the pupils are channelled into different streams (scientific, technical, vocational, agricultural, commercial, etc.).

Through guidance and counselling activities young people are acquainted with the various types of education and training programmes, the world of work and job opportunities. The curriculum is so designed as to provide sound general education linked with the needs of local communities or regions. Practical work programmes are centred around simple workshop practices, farming operations and house management.

The underlying objective of this training is the development of better concepts, to understand the relationship of cause and effects, muscular control, neatness, accuracy, professional pride, and above all, a high sense of dignity of labour.

SECONDARY EDUCATION

The next stage is the secondary level, which is at present of two years' duration with ninth and tenth classes, but is to be a four-year stage in future, according to the National Education Policy announcement. The entry to higher education is governed by many factors such as the enrolment capacity of education institutions, the national needs for certain types of manpower and the individual ability to benefit from the programme. A large number of skilled workers and technical personnel for various occupations stream out through this stage of education.

HIGHER EDUCATION

This consists of colleges of arts, sciences, technology or commerce and universities. The entry into these is governed by the ability of individuals and by national needs. The curriculum in the institutions is designed to meet the challenges faced by society. The planning and development of university education and of the technical institutes is discussed separately in this study.

Polytechnic institutes offer three-year courses after ten years of schooling in a wide variety of technical subjects such as automobile and diesel, civil, electrical, mechanical, chemical, refrigeration and air conditioning and computer engineering. The provision of technical education up to the diploma standard in the various branches of engineering and technology depend upon the type of industries existing and contemplated in the particular region. The type of training is not general in character, but of a specific type so that the students are prepared for a particular level of employment in industry. After entry the standard of studies is gradually raised during the course work. Students completing secondary education from technical schools are also admitted to these courses until a sufficient number of technical high schools can be established.

The specialized training of our engineers with a more scientific bias has created a shortage of engineers in the field of production control and administration. In fact, such jobs need adequate practical training along with technical knowledge and the managerial skills required to deal with the labour force. The duties of production engineers in factories are typical examples of such positions. Polytechnic education, which develops adequate skills followed by advanced theoretical and practical training, can easily provide this type of manpower.

Some of the polytechnics have developed into colleges of technology that offer a four-year programme leading to the B.Tech. degree. A polytechnic diploma-holder, after one year of experience in industry, is eligible to seek admission into a B.Tech. pass course. He has to serve for one further year in industry to qualify for entry into the one-year B.Tech. honours course. This degree enables him to call himself a technologist. This category of manpower forms a connecting link between technicians and engineers. At present, engineering-college graduates are performing these duties. The structure of technological and engineering education in Pakistan is outlined in Figure 1.

The Faculties of Engineering in engineering universities have four-year undergraduate engineering courses leading to bachelor degrees in engineering. The admissions to engineering courses are based on merit as exhibited in the intermediate examination after the two years of science

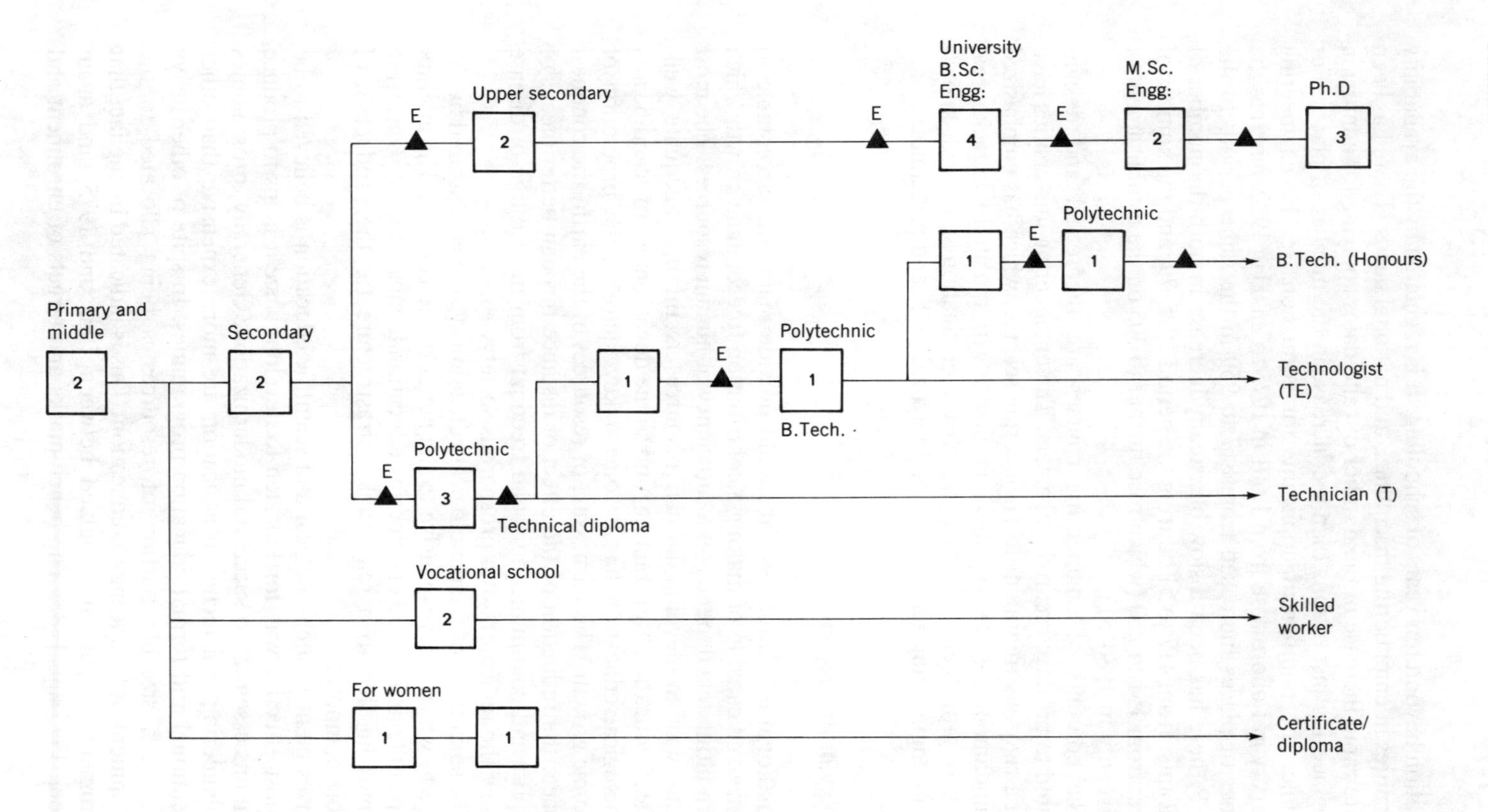

FIG. 1. Technological and engineering education in Pakistan. (Figures indicate years of training; E = entrance examination.)

education beyond ten years of schooling. It is regulated by the availability of facilities in engineering disciplines, and national needs. There has been a more rapid increase in the number of candidates applying for admission in engineering institutions than the increase in the number of places. For example, the number of applicants for admissions to the engineering university in Lahore rose from 1,601 in 1975 to 2,800 in 1980, whereas the number of places increased from 817 to 959 in the same period. In the NWFP there has been a slow but steady increase in both the number of applicants, from 417 to 512 during the period 1975–79, and the number of places, from 151 to 260 (which rose further to 305 on the establishment of a university in 1980–81).

The contents of engineering courses are theoretical and design-oriented to produce creative engineers. The engineering universities have given a rigorous treatment to engineering science, which has emphasized the fundamentals so as to provide graduates with the ability to tackle real engineering problems. Students reach the graduation stage with knowledge of engineering science and of analytical tools and principles.

Strategy of development

The national education system should provide manpower acceptable to industry and capable of making a contribution to economic growth which in turn ultimately depends on a sufficiency of human resources. The most obvious way to develop human resources is through adequate well-balanced education and training, and so the development of scientific and technological education has become a key factor in the promotion of economic growth. The harnessing of resources of the earth, the conquest of space, the reduction of the effect of distance through better means of transport and communication, and freedom from the scourge of disease are now the indicators of a prosperous society.

The soundness of a technological education system is measured by the degree to which it is responsive to the needs of the country. The attempt to expand technical and engineering education to meet Pakistan's growing demands has led to a reasonable infrastructure for the production of suitable manpower.

Investment in technological and scientific education is believed to be the most effective way for Pakistan to develop, since it helps to produce conditions essential to successful modernization. Generally, programmes of engineering and technical education are more expensive than other conventional and formal education programmes. In spite of other heavy pressures, a sizeable portion of resources is being allocated to the development of technological education. This is indicated by the fact that government expenditure doubled between 1965 and 1975, and again between 1975 and 1983. The bench-marks and targets of enrolment and output as envisaged in the Fifth (1978–83) and the Sixth (1983–88) Five-Year Plans are indicators of the facilities developed and planned.

In the early stages of development, the main aim was to accelerate the expansion of the engineering, technical and vocational education programme and this was given high priority. A stage has now been reached when it is proper to concentrate on the quality of education. This depends on critical factors such as manpower planning, educational structure, curriculum design, staff development, student evaluation and systems of examination, technical-learning resources, and educational research and innovation.

MANPOWER PLANNING

The development of education plans based on assessment of manpower requirements is a complex problem because reliable statistics are not available. Therefore, the method of estimating manpower needs has become an important problem. Generally, the following methods are used:

1. On the basis of the relationship between the productive capacity of different units and their expected pattern of employment. Estimates are then made of the total demand for various categories of manpower during the following five to ten years.
2. On the basis of the financial investment needed for development, and the technical personnel needed. For a given size of investment the number of occupational competencies are worked out and translated into various categories of manpower.
3. On the basis of various inputs needed to sustain an economic growth rate, technical manpower being one of the main inputs for securing economic growth.

Of the above three methods, the 'investment and employment' method is the simplest, but it has major limitations. The estimates are likely to be affected due to the changes in the financial policies which also affect the total size of the investment in human resources. Also, technological progress results in a change in the requirements of operational skills. Fluctuations in capital costs as well as operating costs of a scheme will produce errors in the estimates of manpower to be needed over a period of years.

The evolution of engineering education in Pakistan

In 1947 Pakistan inherited a system of engineering education from British India aimed at training engineers to staff public services such as the public works department, railways or irrigation works. A good deal of the work was of a maintenance character which required technicians and few professional engineers of a creative nature. Research and development in engineering fields had hardly begun in Pakistan, yet it is in this area that standards of training and education had to compare favourably with the rest of the world if the country was to industrialize successfully. Research and development had to be important activities and there were needed engineers with vision and talent and with the opportunity to use their skills.

Engineering education and educators were needed to produce engineers of high quality capable of meeting the needs of the nation and of introducing advanced technologies. In 1947 the economic and industrial needs of the country were apparent but the creative engineers available to solve the problems were very few. A considerable number of challenges lay ahead in the area of engineering education.

ENGINEERING EDUCATION AT INDEPENDENCE

There were only two engineering colleges in Pakistan at the time of Independence in 1947. One was at Lahore, affiliated to the University of Punjab, and the other at Karachi, affiliated to the university. Soon after Independence a new engineering college was established at Dacca. The colleges at Lahore, Karachi and Dacca offered, respectively, three, three-and-a-half and four-year degree courses in civil, electrical and mechanical engineering; these changed to uniform four-year degree programmes in 1961. The institutions also offered three-year diplomas in engineering courses but these were discontinued in 1958 after the establishment of polytechnics. Under the pre-1947 British regime, engineering education at the degree level had been controlled by the government and related rather closely to the apparent need for engineers.

The prerequisite for entry to the degree course remains the same as prescribed in early days of Independence and is the successful completion of the Upper Secondary Certificate examination, with mathematics, physics, chemistry and English as the main subjects. This examination is held after twelve years of schooling, when candidates are about 16–18 years of age.

GROWTH AFTER INDEPENDENCE

The landmarks in the growth and development of engineering education were the recommendations of the Commission on National Education (1959), which set the direction and pace of engineering education in Pakistan. These laid out the blueprint for future development in engineering.

The following is a summary of the recommendations on engineering education:

I. A clear distinction should be made between the function of an engineer and that of a technician, and education programmes for each should be developed accordingly.

II. *Objectives*
The objectives of engineering education should be as follows:
(a) to give the students a competence in applying the principles of mathematical and physical sciences to the solution of engineering problems;
(b) to inspire students with a determination to use local raw materials and to develop new techniques appropriate to our conditions;
(c) to educate students in a sympathetic understanding of the economic and social conditions and ways of life in our country;
(d) to develop in students a creative and imaginative approach to their chosen profession, a strong professional consciousness, a profound sense of personal honour and integrity, and the qualities of community leadership.

III. *Admission requirements*
Qualification for admission should continue to be Intermediate Science. Aptitude tests adapted to local conditions should be experimentally developed to supplement examination results.

IV. *Duration of the course*
1. In all engineering colleges the minimum duration of the degree course should be four years.

2. The diploma courses now being administered by engineering colleges should be transferred to polytechnics and technical institutes.

V. *Types of courses*
1. In addition to civil, mechanical and electrical engineering now taught in the engineering colleges, chemical and mining engineering should also be included. Further new courses should be introduced in fields of great promise for Pakistan, such as metallurgy, mineralogy, ceramics, petroleum, and particularly those which deal with the exploitation of local resources.

2. Curricula should include courses in social studies and the humanities to the extent of 15 per cent of total subject-matter to develop in the engineering student an understanding of social and economic forces. They should also include a series of lectures on professional ethics.

3. Practical training in the field and in industries should be arranged for students during summer vacations and satisfactory performance should be a prerequisite to the award of degrees.

VI. *Examinations*
1. The examination should be an assessment of the student's ability and achievement and not of his memory and data-retention capacity.

2. The examination system should be reorganized and the award of degrees should be based on the performance of the student in (a) the final comprehensive examination conducted by the university (75 per cent), and (b) his record in periodical tests and classwork (25 per cent). Students must obtain pass marks in (a) and (b).

VII. *Postgraduate instruction*
Postgraduate courses are an essential element of higher education in engineering as in other disciplines. These should be started in selected engineering colleges as soon as possible.

VIII. *Research*
Engineering colleges should undertake research in projects which may lie on the borderline between fundamental and applied research, in industrial processes, and in construction and in the materials used in them.

IX. *Facilities*
1. Arrangements should be made to remove the existing deficiencies in laboratory equipment.
2. Departmental heads should be made responsible for laboratory upkeep.

X. *Extension courses*
1. Engineering colleges should organize refresher courses for practising engineers and popular lectures and exhibitions for the general public.
2. Museums of Science and Technology should be established at various centres to arouse public enthusiasm for science, engineering, and industry.

XI. *Teaching staff*
1. The following measures should be adopted to remove the shortage of staff in the engineering colleges:
(a) special efforts should be made to obtain the services of Pakistani personnel working abroad;
(b) a search should be made from among the serving engineers for secondment of suitable ones to the engineering colleges;
(c) vigorous efforts should be made to recruit qualified foreign teaching staff.
2. To attract the best-qualified people as teachers in engineering colleges, the following steps should be taken:
(a) the pay scales, particularly the starting salary, should be raised appreciably and facilities in regard to accommodation and medical benefits provided;
(b) employment procedures should be simplified and the existing rigid limitation in the different categories of staff should be removed, and staff should be appointed at the rank commensurate with their qualifications;
(c) those members of the staff whose work shows marked promise should be sent abroad for higher studies on study leave at an interval of about 5–7 years.

XII. *Assimilation of graduates into employment*
The engineering colleges should appoint a member of the staff as career officer to advise students and alumni on career possibilities.

XIII. *Administration and control*
1. Since teaching and research can best flourish in a university atmosphere and since the engineering colleges are responsible for promoting a programme of teaching and research and for producing personnel who will develop the resources of the country, the colleges should be detached from departmental control.
2. A technical university (or institute of engineering and higher technology) should be established both in East and West Pakistan.

3. The new universities (or institutes of engineering and higher technology) may be established either as entirely new centres or through the expansion or reorganization of some existing colleges. In either case, it will be necessary for such an institution to set up appropriate relationships with existing colleges in order to co-ordinate activities in engineering.

The Government of Pakistan accepted the recommendation of the commission and restructuring of engineering education was carried out to initiate the proposed changes. As the direct result of the recommendations, the Engineering College, Lahore, was upgraded in 1961 into the West Pakistan University of Engineering and Technology, and Ihsanullah Engineering College at Dacca flowered into the University of Engineering and Technology in East Pakistan. In view of the great demand and need for qualified engineers in different fields of engineering, professional or graduate engineers are being educated now in twelve engineering colleges/universities. Facilities for training engineers also exist at the Institute of Chemical Engineering and Technology in the University of the Punjab, the Institute of Textile Technology, Faisalabad, and the Agriculture Engineering Faculties of the agriculture universities at Faisalabad and Tando Jam. For training of marine engineers, a Marine Academy is functioning in Karachi under the Federal Ministry of Communication. The army trains its officers as graduate civil engineers at the Military College of Engineering, Risalpur. The total intake of the institutions is about 3,650 students per year, with an enrolment of 15,333 and an output of 2,400 graduates per year.

The engineering institutions

The leading engineering institutions are as follows: The University of Engineering and Technology, Lahore; The NED University of Engineering and Technology, Karachi; Mehran University of Engineering and Technology, Jamshoro; NWFP University of Engineering and Technology, Peshawar; PAF College of Aeronautical Engineering, Karachi; and the Centre of Excellence in Water Resources Engineering, Lahore.

UNIVERSITY OF ENGINEERING AND TECHNOLOGY, LAHORE

The university's history began in March 1923 when the Government of British India established the Moghalpura Technical College in order to provide engineering technicians for the then Punjab Province. The name was later changed to the MacLagan Engineering College, and again in 1956 to the Punjab College of Engineering and Technology. Finally, following recommendations of a National Commission on Education, the Government of Pakistan upgraded the college to a university in 1961.

New disciplines such as architecture, city and regional planning, and chemical engineering were introduced and a general expansion began. Other degree courses were started in the 1965/66 and 1969/70 sessions so that today the university has two major faculties.

The Faculty of Engineering has nine departments: civil engineering, mechanical engineering, electrical engineering, mining engineering, chemical engineering, metallurgical engineering, petroleum and gas engineering, physics, and the university workshops.

The Faculty of Architecture and Planning has six departments: architecture, city and regional planning, mathematics, chemistry, Islamic studies, and humanities and social sciences.

The undergraduate courses lead to the B.Sc. degree in various branches of engineering and in architecture, and in city and regional planning. There are also postgraduate courses leading to the M.Sc. degree in civil engineering (five specialisms), mechanical engineering (three specialisms) and electrical engineering (four specialisms). The M.Sc. can be obtained by full-time or part-time study lasting at least one and two years respectively.

Research facilities are such that since 1975 students have been accepted as candidates for the Ph.D. degree in civil, mechanical, electrical and chemical engineering. Research activities were expanded soon after 1961 following the establishment of a Directorate of Research, Extension and Advisory Services at the university, under the control of the Committee for Advanced Studies and Research. The directorate initiates research activities and disseminates the results of research. It also assists the departments in these activities and in extension and advisory services and also co-ordinates research work with that in industry. The policy of the university is to help industry solve its problems wherever possible.

Admission to the undergraduate programme is strictly on merit and is based on the results of the F.Sc. and pre-engineering examinations or in grades obtained by holders of a B.Sc. in mathematics and physics. The annual intake is 800 students (650 at Lahore and 150 at a constituent college at Tasula). Out of these, civil, mechanical and electrical courses each absorb 200 students with the remaining 200 distributed amongst six other disciplines. The university basically serves Punjab Province but students from other provinces are admitted on a reciprocal basis, the students being selected by the various provincial governments.

Candidates from foreign countries must be sponsored by their governments and today there are 210 such students coming from the Islamic Republic of Iran, Jordan, Kenya, Kuwait, Nepal, Qatar, Saudi Arabia, Sri Lanka, Sudan and the United Arab Emirates.

In 1961 the university had 447 students; in 1983 there were 2,665 undergraduates and 150 postgraduate students. Many of the graduates go to work in Middle East countries and Pakistan has suffered from this loss of qualified engineers. To meet this shortage the university is to be expanded by the creation of four new constituent colleges.

The university also has affiliated colleges such as the College of Textile Technology at Lyallpur, and four technical colleges in Rawalpindi, Lahore, Rasul and Sialkot. The new B.Tech. programmes at these colleges will have their standards and examinations moderated by the university.

The university has a beautiful campus in Lahore and the buildings comprise not only those for the academic departments but also several residential halls for 2,000 students. These halls are equipped with modern amenities. Another hall is now under construction. The halls have some recreational facilities but there are also a number of sports grounds and a new sports centre is now being built. There is also a new swimming pool.

There is residential accommodation for teachers and other staff on the campus itself, to deliberately foster informal contacts between staff and students. There is, in fact, a Bureau for Counselling and Guidance whose activities also complement those of the departments and is concerned with many aspects of student guidance, including career planning.

There is a practically free health service, with four doctors and many para-medical staff, for students and university employees.

NED UNIVERSITY OF ENGINEERING AND TECHNOLOGY, KARACHI

This university is named after Nadirshous Edulgi Dinshaw who provided funds in 1922 in order to set up the oldest engineering institution in what now is Pakistan. It began as a civil engineering college with 50 students, but now has 2,400 undergraduate students and 80 postgraduates (in civil engineering). The college outgrew its old campus and a new one was acquired alongside the University of Karachi, building plans being approved in 1967. With World Bank assistance, Phase I of the plan was completed in 1975 and some 1,500 students are on the new site. Phase II of the development plan has started and, when completed, will enable at least 2,000 students to work on this site. There are offices, classrooms, laboratories and workshops and also an air-conditioned auditorium, a library with 50,000 books, a medical centre, a gymnasium and residential accommodation for about half the total number of students and staff. Sports facilities, a mosque, and further residential accommodation and cafeterias form part of the Phase II development.

There is a staff development scheme enabling junior staff to study for higher degrees abroad. So far, twenty-four students have left Pakistan under this scheme, with twelve already having returned.

There are four colleges in Karachi affiliated to the university as regards academic supervision: the Dawood College of Engineering and Technology (formerly the National College of Engineering), the PAF

College of Aeronautical Engineering, the Pakistan Naval Engineering College, and the Government College of Technology.

On the main campus are the Departments of Civil, Mechanical and Electrical Engineering, the Department of Mathematics and Science and a Department of Humanities. There are B.Eng. courses offered in civil, electrical and mechanical engineering which have a minimum duration of four years, divided into eight semesters. There are theoretical and practical examinations. Since 1979/80 the Department of Civil Engineering has accepted postgraduate students and at the moment the M.Sc. can be awarded in civil engineering for studies in structural engineering, or soil mechanics and foundation engineering, or transportation engineering. Other specialisms will be added in due course. The M.Sc. teaching programme is carried out in the evenings to enable practising engineers to attend courses and to use the services of eminent engineers as teachers. The minimum duration of the M.Sc. course is therefore two years or four semesters.

MEHRAN UNIVERSITY OF ENGINEERING AND TECHNOLOGY, JAMSHORO

This university in Sind Province has two campuses: one at Jamshoro and the other at Nawabshah. At the time of Independence the NED College of Engineering in Karachi was the only engineering college in Sind. As the demand for engineers increased the NED College, with limited facilities, had to restrict admissions to students from Karachi. However, the majority of the Sind population lives in rural areas with incomes well below the national average. To provide opportunities for people in these areas to acquire engineering qualifications and participate in national development, and perhaps improve conditions in rural Sind, another engineering college, the Sind University Engineering College, was set up in 1963 as a constituent college of the University of Sind, Jamshoro.

In 1972 the government decided to upgrade Sind College to the University of Engineering and Technology situated at Nawabshah and, in new premises there, first-year students were admitted in 1974. However, second- , third- and fourth-year engineering classes continued in Jamshoro in the original Sind College premises. Today first- and second-year students are taught at Nawabshah and third- and fourth-year students at Jamshoro. The two campuses now form the Mehran University of Engineering and Technology, created in 1977, but having passed through an intermediate stage of being part of the University of Sind (1976/77). The seat of the new university is at Jamshoro with the college at Nawabshah being a constituent college.

From the 1979/80 session onwards first-year students have been admitted at Jamshoro and Nawabshah. B.Eng. degree courses are now given at Nawabshah and Jamshoro in civil, mechanical and electrical engineering. At Jamshoro, however, there are also B.Eng. courses in chemical engineering, mining and metallurgical engineering, industrial engineering, and architecture and town planning. B.Eng. courses are of four years' (eight semesters') duration, except in architecture which lasts five years (ten semesters) and leads to a B.Arch.

There are nine teaching departments: chemical, civil, mechanical, electrical, electronic and mining and metallurgical engineering; architecture and town planning; and basic sciences and related studies.

The campus at Jamshoro includes the administration, the main library and residential accommodation for students and staff. Postgraduate studies also take place at Jamshoro, and postgraduate diplomas and M.Eng. degrees can be awarded in six fields: structural engineering, hydraulics and irrigation engineering, public health engineering, electrical communication engineering, mechanical engineering, and chemical engineering.

The M.Eng. degree can be obtained by full-time study and research (over four semesters) or by part-time study and research (over six semesters) and the degree is awarded after the successful completion of certain courses and the presentation of an acceptable thesis on project work. The post-graduate diploma can be awarded on course work alone.

NWFP UNIVERSITY OF ENGINEERING AND TECHNOLOGY, PESHAWAR

In 1952 the late Khan Abdul Qayum Khan, the then Chief Minister of North-West Frontier Province, laid the foundation stone of an engineering college on what was the campus of the University of Peshawar. Twenty students were admitted in 1952 to study for B.Eng. degrees in electrical engineering and mechanical engineering. The following year a Department of Civil Engineering was set up. The engineering college was in fact the Faculty of Engineering of the University of Peshawar. In 1980 the college of engineering became a fully fledged university. The student intake is now 340 students per year and by 1984 some 3,810 graduates had passed from the institution.

The university has five degree-awarding departments: civil engineering, electrical engineering, mechanical engineering, agricultural engineering, and mining engineering. The courses for the bachelor degree last four years with annual examinations. Project work in the final year is a necessary part of the courses for practical skills are accorded great importance. The civil engineering B.Sc. is not awarded until some 800 hours of practical training in industry has been acquired.

The departments have well-equipped laboratories which are adequate to serve the schemes of study. The Department of Agricultural Engineering is concerned with the production, processing and handling of foods and fibres and their preservation. Special attention is placed on water management. The university is proud of the standard of its courses which are on a par with those in more developed countries.

Admission to courses at the university is by merit and is open to both sexes. The number of places available in the various disciplines is announced each year when applications are being made.

The university has various support facilities such as a major university workshop. This workshop is used not only by the technical staff but also as a teaching unit. All engineering students get a practical training and mechanical engineering students receive more extensive instruction in aspects of production engineering. Another support unit is the Research Cell whose function is to encourage and organize applied research programmes, especially on projects of immediate benefit to the rural population. The Scientific Instrumentation Centre, established in collaboration with the United Nations Development Programme in 1983, has electronic, fine mechanics, and glass-blowing workshops and, amongst other work, repairs instruments for the university, as well as for the University of Peshawar and Khyber Medical College.

Games and sports are looked upon as being of great importance and there is a University Sports Director. Other extracurricular activities such as dramatics, photography, etc. are also catered for.

At present only the Department of Civil Engineering has postgraduate courses. M.Sc. degrees are offered in water resources engineering and structural engineering. The full-time course lasts two years and can be obtained by course work and a project, or by course work and research. The library, with 35,000 books, also subscribes to a series of technical journals, as needed for research and advanced studies.

PAF COLLEGE OF AERONAUTICAL ENGINEERING, KARACHI

This college teaches engineering cadets of the Pakistan Air Force in aerospace and avionics engineering and prepares them for commissions in the maintenance and electrical branches of the service. The courses lead to the degree of B.Eng. of the NED University of Engineering and Technology, Karachi, with which this college is affiliated.

The PAF College is controlled by Air Headquarters, and has five teaching departments: humanities and science; avionics; aerospace; industrial engineering; military science. The courses are not only open to cadets but also to serving officers of PAF and other services, government departmental engineers and to officers from certain foreign countries.

The departments have facilities for both teaching and research. The Department of Aerospace Engineering has the following laboratories: aerodynamics, gas dynamics, propulsion, heat transfer, structures, mechanics and materials science.

The Department of Avionics Engineering has the following laboratories: basic electronics, advanced electronics, computers and control, integrated circuits, microwaves, antennas, communications and radar, and electric machines.

The Department of Industrial Engineering has machine shops, a metrology laboratory, and a production engineering laboratory, whilst the Department of Humanities and Science has physics, chemistry and computing laboratories. There is also a well-stocked College Library of books and journals.

CENTRE OF EXCELLENCE IN WATER RESOURCES ENGINEERING, LAHORE

The functions of this centre, which is affiliated to the University of Engineering and Technology, Lahore, are to: (a) engage in high-level teaching and research; (b) train research workers; (c) establish M.Phil. and Ph.D. programmes; (d) promote co-operation with other teaching and research establishments; (e) arrange conferences, seminars and refresher courses; (f) conduct teaching and research in such subjects as are assigned to it by the Federal Government.

To carry out these functions the centre has new premises on the campus of the University of Engineering and Technology in Lahore. There are the usual offices and a library, and new laboratories for hydrology, irrigation and reclamation. However, these latter are not yet fully equipped and so there is emphasis on computer-oriented research projects. There are IBM 1130 and 370 computers at the centre and several mini-computers.

Model facilities for irrigation structures and flow measuring apparatus are available in the laboratories, and there is apparatus for soil and water analysis both in the laboratory and in the field. Thus at present the centre offers postgraduate courses in water resources management, and hydrology. When more equipment is fitted and staff recruited there will be postgraduate courses introduced in water resources engineering and hydraulic engineering.

These courses are part of the requirements for the M.Phil. degree, the other part being an acceptable thesis on some research topic. Full-time students can complete the requirements in two years, whilst part-time students must attend for at least three years.

The Vice-Chancellor of the university is Chairman of the Board of Governors of the centre, which is headed by a Director who reports directly to the board.

GOVERNMENT POLICIES

Over the years the Government of Pakistan has made generous provision for all universities including the technological universities. The amounts allocated (in rupees) for each of the six Five-Year Plan periods are as follows: First Five-Year Plan (1955–60), 40 million; Second Five-Year Plan (1960–65), 59 million; Third Five-Year Plan (1965–70), 59 million;

TABLE 1. Student enrolment and number of teachers in Pakistan's technological universities, 1983/84

University	Enrolment 1983/84	Intake (1983)	Teachers (1983/84)
Mehran University of Engineering and Technology	2 405	600	168
NED University of Engineering and Technology	2 978	627	85
NWFP University of Engineering and Technology	1 120	340	89
University of Engineering and Technology, Lahore	3 821	1 104	322

Fourth period (non-plan) (1970–78), 399 million; Fifth Five-Year Plan (1978–83), 687 million; Sixth Five-Year Plan (1983–88), 2,000 million.

During these years the enrolment of students and the number of teachers in the four technological universities has steadily increased. (See Table 1.)

The country is passing through a phase of rapid technological development and needs more engineers in the foreseeable future. The existing engineering universities will soon reach optimum size, and more engineering colleges will have to be established. These colleges are needed not only because of student pressure but because Pakistan is characterized by population growth, a low per capita income, unequal distribution of wealth, a low rate of saving and investment, and a dependence on a few primary products. Technological development is essential if a healthy balanced economy is to be achieved. This will depend on a sufficient number of educated and trained people including engineers and technologists.

In the Sixth Five-Year Plan it was noted that the present teaching programmes were largely at the B.Sc. level, with only limited progress in the development of M.Sc.-level engineering programmes. The B.Sc. programmes varied in quality and growth and had to be restricted so as not to out-run the capabilities and equipment of the existing institutions. Stronger links were urged between the engineering colleges and industries, and eminent Pakistani engineers working abroad encouraged to return as visiting professors.

Projects in existing engineering universities and colleges should be completed, and in view of the requirements foreseen by the Seventh Plan, a new engineering university should be set up in Punjab.

Control and structure of the universities

UNIVERSITY GRANTS COMMISSION

The universities now receive funding from the Federal Government. The link between the government and the universities is the University Grants

Commission, established in 1973 and given a charter in 1974 by an Act of Parliament. Its primary purposes are to make objective assessments of the requirements of the universities and to secure adequate funds to satisfy these needs. The commission is responsible for disbursing recurrent and capital grants to the universities, and also provides funds for approved projects such as research and sports and cultural activities. It has also provided for junior and senior fellowships and instituted chairs and posts for visiting professors. The commission maintains an information service on higher education for government departments and other interested bodies and individuals. It maintains a close and continuing dialogue with the government, the Vice-Chancellors of the universities, teachers, scientists and educationists. It is thus an effective forum for consultations, discussions and plans concerning the advance of higher education. It thereby helps to co-ordinate developments in the various universities and so involve them more closely with national plans for socio-economic development.

To do its work the commission may:

Enquire into the financial needs of the universities and prepare quinquennial programmes for their development.

Allocate and disburse grants to the universities for approved projects and ensure the proper utilization of such grants.

Receive schemes and requests from the universities and, after scrutiny, recommend them to the Federal Government or a provincial government for grants-in-aid.

Collect information and data on all such matters relating to university education in Pakistan and other countries as it thinks fit, and make the same available to the Federal Government or a provincial government, universities and any other interested agencies.

Institute fellowships, scholarships and visiting professorships in universities.

Support and co-ordinate the research programmes of the universities.

Supervise generally the academic programme and development of various institutions of higher learning and education.

Recommend to the universities the measures necessary for the improvement of university education.

Perform such other functions, not inconsistent with the provisions of the Act, as may be consequential to the discharging of the aforesaid functions.

Bring to the notice of the Federal Government or a provincial government the problems of teachers and students and recommend measures for solution thereof.

After consultation with the Federal Government or a provincial government and a university, cause a visitation to any department of the university in such manner as may be prescribed, by a person appointed by the Commission.

After such visitation communicate views to the Federal Government or provincial government, and the university concerned, together with any recommendations regarding any action to be taken.

PAKISTAN ENGINEERING COUNCIL

The Pakistan Engineering Council was established under Act of Parliament in January 1976 and makes provision for the regulation of the engineering profession and oversees the quality of engineering education in Pakistan. The council has power to require information as to the courses of study and examination. It appoints inspectors to all the examinations held by the engineering institutions in Pakistan for the purpose of granting engineering qualifications or in respect of which recognition has been sought. The inspectors report to the council on the adequacy of the examination which they attend and on the courses of study and facilities for teaching provided by the institution in question and on any other matter the council may require them to report.

The council has the power to withdraw recognition of an engineering qualification of an educational institution when in its considered opinion the courses of study and examinations fall short of the standard of proficiency required from candidates holding such qualifications.

UNIVERSITY STRUCTURE

The control of the university also extends to the institution through its titular head, the Chancellor. In the Federal area the President of Pakistan is the Chancellor of a university, whilst in a province the Chancellor of each university is the Governor of the province. The Chancellor can therefore make a significant contribution to the welfare and progress of the university and play a very positive role in its administration. In many ways each university is an autonomous institution with its own policies. The chief administrator is the Vice-Chancellor who is appointed by the Chancellor to serve for several years.

The Vice-Chancellor is assisted by the Syndicate, a compact body of people whose chairman is the Vice-Chancellor, and is responsible for budget estimates and financial matters, for its property, and for the appointment of certain personnel. The members of the Syndicate are appointed by statute for their competence in these affairs.

The Vice-Chancellor is also chairman of the Academic Council which is responsible for academic affairs such as approval of teaching programmes, academic standards, examinations, award of degrees and so on. The council is composed of professors, heads of departments and senior teachers, the principals of affiliated colleges, and external members.

The day-to-day business of a university is carried out by its departments and administrative officers advised by various Boards of Studies and Special Committees. Each Board of Studies will be responsible for the preparation of schemes of studies and syllabuses in some discipline. Special Committees will be concerned with finance, or planning, selection of teachers, discipline, and so on.

The quality of teaching and research depends on the quality of the academic staff. They are appointed into four categories:

Lecturers. First-class B.Eng. or equivalent qualification required.

Assistant Professor. Lecturer qualification, plus at least six years' teaching/research experience in a university or postgraduate institution or recognized organization, *or* plus M.Sc. in engineering and four years' acceptable teaching or research experience, *or* plus Ph.D. and two years' teaching or research experience.

Associate Professor. Master's degree plus at least thirteen years of teaching/research experience in appropriate institutions, *or* Ph.D. and ten years' suitable experience.

Professor. Master's degree plus at least eighteen years of teaching/research experience, *or* Ph.D. and fifteen years of suitable experience.

With time the requisite qualifications necessary for each post are tending to rise as more qualified people are available for university posts. All the universities have staff development programmes enabling junior staff to move up the promotion ladder.

EDUCATIONAL POLICY

Entrance to an engineering course at a technological university depends first of all on successful results in the Upper Secondary Certificate examination, or equivalent, with the main subjects being mathematics, physics, chemistry and English. The examination is held after twelve years' schooling, so the candidate's age is sixteen years and above. The number of applicants far exceed available places, so a solution has to be found. Different institutions adopt their own procedures; some not only give weight to performance in the Upper Secondary Certificate examination but hold their own entrance examination; others use special engineering aptitude tests, or school records etc. But in the end only the best students are admitted.

The policy in all the universities is to provide a rigorous treatment of engineering science subjects which emphasizes fundamentals so as to provide the graduates with an ability to master real engineering problems when in employment. Students usually graduate with a knowledge of engineering science and analytical methods but with little experience in the application of their knowledge to engineering tasks. This skill has to be acquired after graduation. There are no formal sandwich or co-operative degree courses in Pakistan.

Engineering curricula are designed so that the time spent on various topics is generally as follows: humanities (Islamic studies, economics, management, report-writing, etc.), 15 per cent; engineering science (including mathematics, physics and chemistry), 25 per cent; general engineering (practical engineering topics over a wide field), 20 per cent; specialized engineering (major field of engineering), 25 per cent; current

TABLE 2. Distribution of academic staff, 1983/84

University	Professors	Associate professors	Assistant professors	Lecturers
Mehran University of Engineering and Technology	23	30	47	71
NED University of Engineering and Technology	15	18	21	39
NWFP University of Engineering and Technology	11	22	24	32
University of Engineering and Technology, Lahore	70	86	80	89

national problems (national development and its problems), 5 per cent; engineering practice and applications, 10 per cent.

In Pakistan today bachelor degrees in engineering can be obtained with a major emphasis or specialization in the following fields: civil engineering, architecture, town planning, agricultural engineering, mechanical engineering (and nuclear engineering), metallurgical engineering, mining engineering, textile engineering, chemical engineering, marine engineering, electrical engineering, electronic engineering, tele-communications engineering, aeronautical engineering and avionics.

As industrialization increases so more specialisms will be offered. Nevertheless, every course has a general content for the technological future and progress is uncertain. For example, considerable attention is given to mathematics and its uses throughout each year of the four-year courses, and of course to the use of computers and systems analysis.

In 1983/84 the academic staff distribution was as shown in Table 2.

Postgraduate studies

The postgraduate courses available in Pakistan are as follows:

1. Civil engineering: (a) structural engineering; (b) soil mechanics; (c) hydraulics engineering; (d) highways and transportation engineering.
2. Public health engineering.
3. Town planning.
4. Electrical engineering: (a) power engineering; (b) electronics engineering; (c) computer engineering.
5. Mechanical engineering.
6. Nuclear engineering.
7. Chemical engineering.
8. Water resources engineering.
9. Industrial engineering.

In view of the limited resources available at present, especially in respect of qualified staff, the key postgraduate courses which are recognized as being of major importance to national development are to be concentrated in a limited number of Centres of Excellence to be developed on correct lines. The establishment of the Centre of Excellence in Water Resources Engineering at Lahore is a step in that direction. In addition to the centres' developing postgraduate programmes, they have the responsibility of establishing co-operative relationships with other technical institutions to help in solving the problems in their areas of specialization. Special courses are planned in fields of immediate practical value.

RESEARCH AND DEVELOPMENT EFFORTS

Engineering research is seen as vital to the economic development of Pakistan. New materials which can be produced locally are urgently needed, as well as new techniques which will make efficient and appropriate use of the very large amount of available manpower. Little research is done by private industry, so the responsibility for research and development falls on government and on the engineering institutions.

There are two councils responsible for the promotion and support of research. They are the Irrigation, Drainage and Flood Control Research Council, and the Works and Housing Research Council. In addition, there are the National Institute for Power, the National Institute for Electronics, and the National Institute for Silicon Technology, which carry out research projects and also support research in the universities. In the universities the majority of the academic staff are concerned with teaching and have little time indeed for research. It is only in the Centres of Excellence associated with the universities where most teaching is at the postgraduate level and where the staff have adequate time for research.

The Sixth Five-Year Plan (1983–88) has provided for a much greater input into a much larger variety of research and development programmes than hitherto. The financial provision for technological research is 7 billion rupees in the Sixth Plan, as compared with 2 billion in the Fifth Plan. There will be an improvement in the provision of postgraduate studies at the universities. For example, the Centre of Excellence in Water Resources Engineering at Lahore will be further strengthened so that the quality of the M.Phil. and Ph.D. programmes will be raised. In future, postgraduate student numbers are likely to be in the region of 4 to 10 per cent of the university student population.

Future trends

During the present Five-Year Plan the spectrum of undergraduate engineering studies is being broadened, and there is an interesting new development aimed at improving teaching. Selected departments at the engineering universities will be developed into, or have associated with them, Centres of Academic Studies. These centres will be equipped and staffed so that in collaboration with well-known foreign universities they can develop programmes in certain advanced studies. The subjects identified for advanced study programmes at certain universities are as follows:

University of Engineering and Technology, Lahore: public health engineering; materials and building science; soil mechanics; structural engineering; town planning and architecture; production engineering; power engineering; communication engineering.

NWFP University of Engineering and Technology, Peshawar: water resources engineering; structural engineering; power engineering; rural development engineering; refrigeration and air-conditioning.

Mehran University of Engineering and Technology: irrigation and drainage engineering; transportation engineering; industrial engineering; electronic engineering.

NED University of Engineering and Technology, Karachi: structural engineering; environmental engineering; automobile engineering; metallurgical engineering; avionics and aerospace engineering.

For the purposes of estimating student admissions to the engineering universities it is assumed that the population will go on increasing at 3 per cent per annum, but that the admissions of students to undergraduate courses can increase by 4 per cent per annum. Thus enrolments are expected to increase as shown in Table 3.

Admission to postgraduate courses and research will remain limited for some years and the expected trend of enrolment is given in Table 4.

TABLE 3. Projected undergraduate admissions to engineering universities

University	1982/83	1983/84	1988/89	1993/94	2000/01
Mehran University of Engineering and Technology	500	550	650	700	850
NED University of Engineering and Technology	500	550	600	600	600
NWFP University of Engineering and Technology	300	300	400	500	600
University of Engineering and Technology, Lahore	1 100	1 200	1 500	2 000	2 600
TOTAL	2 400	2 600	3 150	3 800	4 650

TABLE 4. Projected postgraduate admissions to engineering universities

University	1982/83	1983/84	1988/89	1993/94	2000/01
Mehran University of Engineering and Technology	20	20	50	80	125
NED University of Engineering and Technology	20	20	30	50	70
NWFP University of Engineering and Technology	10	10	20	30	50
University of Engineering and Technology, Lahore	55	60	75	200	300
TOTAL	105	110	175	360	545

CONTINUING EDUCATION

The need for the continuing education and re-training of engineers and technologists is well-recognized and is already practised in Pakistan, but will be strengthened in the future as resources become available. The following opportunities exist and are practised to some extent now but these activities will certainly become more important in years to come:

1. Use of part-time teachers drawn from industry.
2. Use of teachers as consultants to industry.
3. Temporary exchange of engineers between colleges and industry.
4. Release of engineers to undertake full-time, part-time, or short-term courses on advanced subjects at the universities.
5. Participation of industry-based engineers in university seminars and training workshops.
6. In-house training by industry, perhaps drawing in outside teachers.
7. Using the facilities of international and regional agencies. (In fact, such facilities have been used only sparingly, possibly because of long delays in getting applications processed.)

The evolution of technical education in Pakistan

The progress of technical education, which is the education of technicians and technologists who are not professional engineers, since 1947 is reflected in various steps taken as a result of national policy decisions, usually as part of the five-year development programmes. The period since Independence can be divided into four distinct phases, as outlined below.

PHASE ONE: POLICY FORMATION AND IMPLEMENTATION, 1947–59

Technical and vocational education in the new state of Pakistan was such that it was necessary to state in precise terms the national objectives, or goals, and to set up a framework within which plans could be implemented. A major step was the First Education Conference in November 1947 which considered socio-economic and scientific aspects of education. The founding father of Pakistan, Mohammed Ali Jinnah, in a message to the conference, emphasized the need for technical education as follows:

> Education does not mean academic education only. There is an urgent and immediate need to give scientific and technical education to our people in order to build our future economic life and to well-planned industries particularly. We should not forget that we have to compete with a world which is moving fast in these directions.

A Committee of Technical Education was therefore set up in 1948 which prepared a comprehensive plan for the education and training of various categories of manpower, such as factory executives, technical specialists, supervisory technical staff, and skilled technicians, workers and entrepreneurs.

Limited activities had been carried on during 1947 in two engineering colleges, one in Karachi and one in Lahore. Licentiateship courses for supervisors were offered in civil, mechanical and electrical engineering subjects to supervisory staff needed to maintain the public utility services.

In 1951 the First Six-year Development Plan, 1951–57, was begun and during this period the first polytechnic was created in Karachi to train engineering technicians. In 1958 the (Swedish) Pakistani Institute of Technology was formed at Landhi, Karachi, to train practically biased 'industrial technicians'.

The First Five-Year Plan, 1955–60, allocated 278 million rupees for technical education and engineering education. The Ford Foundation made a survey of technical education requirements and a second polytechnic was created at Rawalpindi. The polytechnics were modelled on American institutions for the education of higher technicians and technologists. At the technical institutes, courses were formed in subjects such as foundry work, pattern-making, welding and so on.

During this Phase One education and training programmes in power engineering, automobile engineering and radio and electronics were introduced for the first time to meet the demands of expanding industries. The transfer of technical education from the traditional engineering colleges to new polytechnics and technical institutes gave a new status to technical education.

PHASE TWO: EXPANSION AND CONSOLIDATION, 1960–69

This phase began at the same time as the Second Five-year Development Plan, 1960–65. The allocation of resources in this plan was based primarily on requirements and estimates made in a National Education Commission report. This commission had made many recommendations, including the following on technical and vocational education: (a) a network of vocational institutes should be established to train craftsmen; (b) more technical institutes should be established with the aim of producing 7,000 technicians by the end of the Second Five-Year Development Plan; (c) all polytechnics should offer part-time evening courses to suit the needs of employees of small firms and local industries; (d) technical-teacher training wings should be established.

In 1964 there were established Regional Directorates of Technical Education for the administration and co-ordination of the new institutes and polytechnics. There was a main board at Lahore which had jurisdiction over all institutes in West Pakistan, and indeed had control of academic activities of all vocational, commercial and technical institutes.

A large number of technical institutes were expanded and up-graded to polytechnics. Demand for education was such that private technical institutes emerged. Commercial education was greatly expanded, and facilities for the vocational education and training of women were reorganized and strengthened. Indeed, by the end of the Second Five-Year Plan period in 1965 the progress and expansion of technical education was seen to have been more rapid than had been estimated. The seven new technical institutes or polytechnics planned had all become operational and admissions totalled 2,200 (200 more than estimated) and the number of diploma-holders 1,275 (275 more than estimated).

At the beginning of this period, in 1960, there were two polytechnics with an annual intake of 912 students and an output of 432 students.

It had been estimated by the National Planning Commission that during the period 1960–65 about 7,000 trained technicians would be needed. In fact the total output from all technical institutions was about 5,500, representing a short-fall of 20 per cent.

The Third Five-year Development Plan, 1965–70, emphasized the consolidation of existing programmes but with the expansion of student numbers in the technical institutes and polytechnics. Nevertheless, many new technological courses were introduced.

It was now planned that the annual intake of students should be raised to 7,000 per annum by 1970, for the total number of technicians required during the plan period was estimated at 12,000. The nine existing polytechnics were expanded and a further twelve polytechnics founded, and it was proposed that by 1970 there should be thirty polytechnics, offering courses in thirty technologies.

In the meantime a double-shift programme was introduced at some polytechnics in order to use facilities to the full. Noteworthy is the fact that one of the new polytechnics was for women only. The enrolment capacity was thus raised to 9,000 students by 1970, with an annual output of about 2,500 diplomates.

The Fourth Five-Year Development Plan, 1970–75, on the basis of the existing Pakistan was abandoned in the turmoil following the splitting off of East Pakistan as the new state of Bangladesh, and a new educational policy was announced in March 1972.

PHASE THREE: INNOVATION AND EXPERIMENT, 1970–77

This period witnessed new approaches to educational planning and implementation as a result of the New Educational Policy document of 1972. Its major recommendations were the following:

Secondary-level technical courses were to be increased, which enhanced the usefulness of existing rural vocational institutes. Also a number of technical high schools were to be set up in most parts of Pakistan.

Bachelor of Technology (B.Tech.) courses were introduced in upgraded technical institutes and polytechnics in order to produce technician-engineers, or technologists, of the required quality.

Vocational subjects would be added to the content of general education in all schools. The rationale for this was 'the expansion of science and technical education will result in a progressive integration of general and technical education in secondary schools and colleges, so that school-leavers with intermediate level certificates should be ready to accept social responsibilities as a middle-class technician or worker within his socio-economic framework'.

Vocational education for women was to be reorganized so that there would be a new type of school at each *taluka*.*

The provision of B.Tech. courses was a new departure and some selected polytechnics were converted into colleges of technology. At the end of the three-year courses normal to a polytechnic, the students would go into industry. After two years in industry the diplomates could take a further one-year course leading to the degree of B.Tech. (pass degree level). There would also be a further advanced one-year course whereby a B.Tech. (Pass) could be upgraded to a B.Tech. (Honours).

Nationalization of private technical institutions was also part of the policy to integrate general and vocational education. The lack of an adequate infrastructure and of resources meant that by 1975 only 1,200 schools had been able to introduce the new integrated courses.

* *Talukas,* and *tehsils* of which they form part, are rural administrative units of districts created for land revenue purposes.—Ed.

PHASE FOUR: IDEOLOGICAL RE-ORIENTATION, 1978 TO DATE

In 1977 educational policy and planning changed again following the National Conference on Education held in October. The new policy set out to do the following: relate education to national goals and ideology; identify and strengthen all institutions and elements of education; link education to productivity and thus to the standard of living; strengthen the ideological foundation of all educational activities, which could lead to changes in secondary and tertiary education; reorganize the polytechnics and their programmes and offer diploma courses in the evening as well as during the day. A new experimental type of village 'workshop' would be tried out as part of this programme.

The present four-tier system (primary, secondary, college and university) will be replaced by a three-tier system. This will be elementary (grades 1–8), secondary (grades 9–12), and university level. The university level includes all institutions offering degree courses (B.A., B.Sc., B.Tech., etc.). This proposal needs further study as there are many implications regarding the number of places needed, and teachers, in secondary institutions. However, the overall aim is to produce enduring self-reliance and thus sustained economic growth.

There was a Fifth Five-Year Plan, 1978–83, which set new targets for the outputs from engineering, technical and vocational institutes (Table 5).

The annual output of skilled workers (craftsmen, tradesmen, etc.) from twenty-three institutes was to be raised from 928 to 2,700 by increasing the enrolment from 2,700 to 7,500. To do this new vocational institutes, or their equivalent, were required.

There are vocational institutes for women and the output (from an enrolment of 4,700 students) was 2,500 in 1977/78; the planned target output in 1982/83 was 3,200.

The Sixth Five-Year Plan, 1973–88 lays great emphasis on extending the provision for technical and vocational education and training. The plan is to increase the number of polytechnics from 28 to 47, and monotechnics from 7 to 17, and there is to be 200 new trade schools and 30 new commercial institutes. Hence the output of polytechnic diplomates

TABLE 5. Targets of the Fifth Five-Year Plan, 1978–83

Type of institution	1977/78	Target 1982/83	Increase (%)
Engineering	1 700	2 360	30
Polytechnic			
B.Tech.	50	390	640
Diploma	3 500	4 925	40
Vocational institutes	928	2 700	197

should rise to 5,000 per annum, and the trade schools produce 4,000 skilled workers each year. However, some 45,000 workers undergo training in the private sector, which is not nationalized. A system of certification of this private training might be introduced, but nevertheless it will be encouraged to expand.

It is suggested also that the trades-school programme requires about 278 schools, or one per *tehsil*, and that technical middle or technical high schools should be introduced in order to train youths who have left school and might want to re-enter the education system. The use of evening classes should be expanded and there should be production units attached to the polytechnics to introduce real practices to students, including those of marketing products.

In the women's polytechnics it is suggested that food preservation, dressmaking, commercial designing and nursing be introduced.

The development of engineering and technical education throughout the various plan periods is shown in Table 6.

TABLE 6. Development of technical and engineering education, 1955–88

	First Plan (1955–60)	Second Plan (1960–65)	Third Plan (1965–67)	Non-Plan Period (1970–78)	Fifth Plan (1978–83)	Sixth Plan target (1983–88)
Technical education						
Engineering universities	—	1	1	3	4	4
Engineering colleges	3	3	5	3	1	3
Polytechnics/monotechnics	4	10	14	28	31	40
Intake capacity						
B.Sc. (Eng.) courses	—	603	1 500	2 621	3 650	5 000
Polytechnics/monotechnics	—	2 100	3 500	5 271	5 950	7 425
Enrolment						
Engineering classes	1 251	1 632	3 800	9 500	12 800	15 500
Diploma classes						
Engineering colleges	773					
Polytechnics/monotechnics	722	5 846	7 800	13 100	17 320	20 950
Annual output						
Engineering graduates	304	453	1 256	1 700	2 360	3 550
Diplomates	432	2 100	2 500	3 500	3 970	4 996
General education						
General universities	3	4	6	10	15	15
Enrolment	1 662	8 319	15 475	—	17 000	18 500

COMMERCIAL EDUCATION

The system of commercial education which caters for commerce, business and non-industrial handicrafts and activities, was not well developed at Independence. Therefore the Ministry of Commerce and Education appointed a Committee on Commercial Education in 1952 to examine the situation. The committee reported that a first step should be to provide a preparatory stage of studies at school for those who intend to go into commerce or business. By the age of about 16 boys and girls at school should have a grounding in such subjects as might be helpful in acquiring and assimilating knowledge of commerce at some later stage. Specialized courses in commercial subjects were visualized as necessary in the future.

In 1959 the Commission on National Education pointed out that some schools were offering a few optional subjects in commerce in the ninth and tenth grades but that few students took them. The commission proposed that schools should offer elective subjects in the fields of agriculture, technology and commerce to be taken in addition to compulsory subjects. Vocational education in commercial subjects was introduced into the eleventh and twelfth grades in schools, under the West Pakistan Directorate of Technical Education. The directorate also had set up twelve commercial institutes by 1960 with curricula which altered as national needs changed. By 1978 the number of such institutes was 48 and the annual intake of students was 2,625 and the output 2,300.

During the Fifth Five-Year Plan, 1978–83, a further four institutes were opened with an annual intake of 2,640, bringing the total annual output to about 4,000. There were also many private institutes in operation, but not everywhere. Hence in the Sixth Five-Year Plan, 1983–88, provision has been made for thirty new commercial institutes, of which two will be for women only.

Technical and vocational education for women

As citizens, women have equal access with men to educational facilities from which they can benefit and so take part in building the national economy. Therefore vocational education for females has been provided in Pakistan since Independence. In about 1935, before Independence, Industrial Schools for Girls were started to enable women to learn useful domestic handicrafts such as sewing, knitting, embroidery, etc., and so become better housewives. These schools continued after Independence.

However, as conditions changed there arose a need to train women in commercially useful handicrafts in order to supplement the family income—often earned by one male member of the family on whom many unemployed females depended. The number of industrial schools where women could learn handicrafts has continually increased from twenty-five under the West Pakistan Directorate of Technical Education to seventy-eight at the end of the Fifth Five-Year Plan in 1983. The number of such institutes and their enrolment (7,050) far exceeds similar institutes for boys.

At the higher levels of technical (vocational) education there are five polytechnic institutes for women; at Peshawar, Lahore, Faisalabad, Karachi and Sukkur. These institutes offer many courses amongst which are commerce (at certificate and diploma levels); dress design and dressmaking technology; commercial photography; civil technology; architecture; radio and electronics.

TEACHER-TRAINING INSTITUTES

Three hundred places for students to become vocational teachers are available in two teacher-training institutes: one in Lahore, Punjab Province, and the other in Hyderabad, Sind Province. Women who have obtained a diploma (two-year course) at a vocational institute can obtain admission to the teacher-training courses. These courses last one year, after which a woman can be appointed a teacher in a school or institute. Students are offered advanced instruction in tailoring and dressmaking, fabric printing, fancy woodwork, leather work and machine knitting. There is also instruction in design methods and in teaching methodology.

Structure and control of technical education

The colleges of technology and polytechnic institutes between them offer a very wide range of courses. Five of the colleges of technology, at Multan, Rasul, Lahore and Rawalpindi in Punjab Province, and in Karachi, Sind Province offer courses leading to both the Bachelor of Technology and to the Diploma of Technology.

The B.Tech. degree can be obtained in the following technologies: mechanical production, automobile, refrigeration and air-conditioning, public health, construction and highway, electrical power, radio/electronics, civil, mechanical, and chemical.

The polytechnic institutes and other colleges of technology offer courses which lead to a certificate or a diploma in a technology. They educate and train two types of technician: the engineering technician and the industrial technician. The technician falls between the skilled craftsman and the engineer, but the engineering technician is most likely to work closely with the engineer while the industrial technician is likely to work with and supervise craftsmen. Curricula have been designed which aim at producing both types of technician.

Until recently students of all technologies have studied a common course in the first year, but a Swedish pattern of education has been used as a model for new curricula. The first year will still be based on mathematics, science, some technical courses, and courses in humanities and social sciences. The practical work in the first year will, however, have a definite relationship with the future field of specialization. The curricula for an industrial technician will differ from that of an engineering technician in the same technology by the attention, or weight, given to the various subjects offered. Thus the two curricula are not fundamentally different but what might be called the centres of gravity lie in different places.

In the first year the practical work is chosen from ten fields of technology: civil; mechanical; foundry work; welding and metallurgy; electrical; radio and television; chemical; auto diesel and farming; refrigeration and air-conditioning; and textile. The scheme of studies in the first year is so designed that it serves the immediate goals of training skilled workers and the ultimate goal of producing two types of technician.

Each year has two 20-week semesters, and 18 weeks of each semester is devoted to instruction and 2 weeks to evaluation and tests. The ratio of theoretical to practical instruction is preferably 10 : 90 but in no case worse than 20 : 80. For example, a typical first semester in the first year of a course in electrical technology will have the following curriculum (figures denote hours per week):

	Theory	*Practical*
English 1	1	0
Mathematics 1	3	0
Science 1 (physics)	1	2
Drawing	0	3
Workshop practice	0	27
Islamic/Pakistani studies	1	0
	6	32

Hence the ratio of theoretical to practical instruction is 6 : 32 or about 16 : 84. Note that the students spend 38 hours per week, in the Institute, during most semesters of the courses leading to a certificate or diploma. English is studied during the first three semesters of a course whilst Islamic/Pakistani studies are part of the curriculum of every semester.

Curricula are developed by a curriculum development bureau in collaboration with the provincial Boards of Technical Education.

AN EXAMPLE OF ACTIVITIES IN A TECHNICIAN COURSE

The following is an analysis of the activities undertaken during the course for an industrial mechanical engineering technician:

1. Writing letters.
2. Selecting materials.
3. Use of material handling techniques.
4. Design of simple structures.
5. Maintenance and repairs of equipment.
6. Sales and servicing.
7. Using handbooks, service manuals and reference materials.
8. Cost estimation and cost analysis.
9. Preparing production non-machine schedules and man-machine requirements.
10. Writing technical reports.
11. Designing simple tools, jigs, fixtures and other production devices.
12. Installation of piping system.
13. Putting design changes into effect.
14. Collection and presentation of information on which managerial decisions are to be based.
15. Inspection of processes in operation.
16. Skills in handling machine tools and hand tools.
17. Analysis and interpretation of test results.
18. Supervising construction and fabrication.
19. Use and care of electrical accessories.
20. Preparing working drawings with skill.
21. Drafting with average skill and proficiency in reading and interpretation of drawings.
22. Installation of machinery.
23. Locating and correcting faults in product/equipment, trouble-shooting.
24. Applying safety rules.
25. Use of industrial engineering techniques—method study, work measurement, etc.
26. Supervising production.
27. Use of measuring and recording instruments.
28. Carrying out acceptance and/or performance tests.
29. Maintaining stores and control of inventories.
30. Management of subordinate staff and/or labour.
31. Giving work instructions, verbal and written.
32. Selection of skilled workers.
33. Training of subordinate staff and/or labour.
34. Shop organization for efficient production.
35. Use of steam, gas, water as sources of energy.
36. Drawing up materials, tools and machinery specifications.
37. Operation and maintenance of foundry shop equipment.
38. Operation and maintenance of welding shop equipment.
39. Operation and maintenance of heat treatment furnaces and equipment.
40. Operation and maintenance of material testing machines and equipment.
41. Application of labour laws.
42. Dealing with trade unions.

BOARDS OF TECHNICAL EDUCATION

These boards function in the three provinces of Sind, Punjab and the NWFP, and not only do they participate in the preparation of curricula but, relative to the institutes and colleges of technology, they have some of the functions of a university administration. For example, they are concerned with examinations, the conduct of research, staff development, etc.

When first set up, the polytechnic institutes, which were few and small, were able to operate a system of internal assessment of students which was reliable and of a comparable standard everywhere. As the numbers of institutes is now large, and with a large academic staff with a variety of backgrounds it was thought necessary to have an element of external examinations. The Boards of Technical Education are in charge of these examinations.

In the NWFP, Punjab Province and Baluchistan, the examinations in the even-numbered semesters are conducted by the boards which also supervise a practical examination. In Sind Province the board examines theory in the fifth and sixth semesters and supervises all practical examinations. All other tests are under internal supervision. Objective tests and continuous assessment are used to grade students.

Some of the colleges of technology have teacher-training wings (or departments) which give instruction in examination techniques and carry out studies of use to the polytechnics and colleges. The boards are also concerned with staff development in the institutes and this subject receives constant attention. The continual introduction of new technologies or new techniques brings to a very sharp focus the need for the development and continuing education of teachers. A teacher at a technical institute or college must have a clear and realistic understanding of the philosophy and purpose of technical education as well as a mastery of relevant science and technology. He should be able to develop new curricula and new schemes of study to meet new needs. Therefore numerous training programmes and scholarships have been arranged for technical teacher development which is cyclic and on-going for the academic staff.

Such development is also related to the need for some diplomates in technology who should receive further theoretical education. Therefore the extra courses were introduced at a few colleges of technology and lead to the Bachelor of Technology degree, and a new class of technician-engineers, which a manpower survey has shown to be needed in Pakistan.

Acknowledgements

Thanks are due to the Chairmen of the technical boards, directors of technical education in Punjab, Sind and the NWFP and registrars of the universities for providing statistical data. Thanks are also due to Jehangir Durrani for assistance given in preparing the manuscript for submission to Unesco.

APPENDIX I

GOVERNMENT COLLEGES OF TECHNOLOGY AND POLYTECHNIC INSTITUTES IN 1985 (WITH 1983 ENROLMENT FIGURES)

Punjab Province

Colleges of technology (all offer Diploma and B.Tech. courses)

Lahore	1 091	Rasul	628
Multan	859	Rawalpindi	585

Polytechnic institutes (all offer diploma courses)

Lahore (for women)	174	Bahawalpur	515
Sailkot	809	Swedish/Pakistani Institute, Gujurat	339
Sarghoda	561	Institute for Printing and Graphic Arts, Lahore	117
Faisalabad	645		
Leiah	428		
Shaiwal	738		

Sind Province

Colleges of technology (all offer diploma courses but only the college in Karachi offers B.Tech. courses as well)

Karachi	1 947	Hyderabad	1 172
Khairpur	808	Habid College, Nawabshah	150

Polytechnic institutes (all offer diploma courses)

Karachi (for women)	277	Sukkur	163
Sukkur (for women)	33	Jacobabad	324
Jamia Institute, Karachi	114	Jinnah Polytechnic Institute, Karachi	247
Soufee Eide Zahabi Institute, Karachi	724	Swedish/Pakistani Institute, Karachi	528

North-West Frontier Province

College of technology (diploma courses only)

Peshawar	930

Polytechnic institutes (diploma courses only)

Peshawar (for women)	—	Nowshara (opened in 1984)	—
Haripur	430	Mingara (to be opened in 1986)	—
D. I. Khan	510		

APPENDIX II

SPECIFIC TECHNOLOGIES TAUGHT IN COLLEGES AND POLYTECHNICS

Punjab Province

College of Technology, Lahore

B.Tech. (Pass and Hons.): (a) Mechanical production; (b) Autos, diesel; (c) Refrigeration and air-conditioning.
Diploma: (a) Autos, diesel; (b) Civil; (c) Electrical; (d) Drafting and designing; (e) Mechanical; (f) Radio and electronics; (g) Refrigeration and air-conditioning.

College of Technology, Rasul

B.Tech. (Pass and Hons.): (a) Public health; (b) Construction and highway.
Diploma: (a) Civil; (b) Architecture.

College of Technology, Rawalpindi

B.Tech. (Pass and Hons.): (a) Electric power; (b) Electronics.
Diploma: (a) Auto diesel; (b) Civil; (c) Foundry and pattern-making; (d) Metallurgy; (e) Machine shop; (f) Electrical; (g) Radio and electronics; (h) Mechanical; (i) Television and architecture; (j) Refrigeration and air-conditioning.

College of Technology, Multan

B.Tech. (Pass and Hons.): Chemical.
Diploma: (a) Chemical; (b) Electrical; (c) Mechanical; (d) Radio and electronics; (e) Civil; (f) Textile.

Polytechnic Institute, Sailkot

Diploma: (a) Electrical; (b) Autos and diesel; (c) Civil; (d) Foundry pattern-making; (e) Mechanical.

Polytechnic Institute, Sargodha

Diploma: (a) Electrical; (b) Mechanical; (c) Instrument; (d) Civil.

Polytechnic Institute, Faisalabad

Diploma: (a) Electrical; (b) Mechanical; (c) Public health; (d) Instrument; (e) Auto farm.

Polytechnic Institute, Leiah

Diploma: (a) Electrical; (b) Mechanical; (c) Civil.

Polytechnic Institute, Sahiwal

Diploma: (a) Electrical; (b) Mechanical; (c) Machine shop; (d) Auto farm; (e) Tool-making and tool design; (f) Civil.

Polytechnic Institute, Bahawalpur

Diploma: (a) Mechanical; (b) Electrical; (c) Civil; (d) Farm machines.

Swedish/Pakistani Institute of Technology

Diploma: (a) Autos and farm; (b) Electrical; (c) Mechanical; (d) Foundry; (e) Instrument; (f) Pattern-making; (g) Welding.

Polytechnic Institute (for women), Lahore

Diploma: (a) Dress-making and dress designing; (b) Television and radio electronics; (c) C. Com. and D. Com.

Polytechnic Institute for Printing and Graphic Arts, Lahore

Diploma: Printing and graphic arts.

Sind Province

College of Technology, Karachi

Diploma: (a) Electrical; (b) Mechanical; (c) Radio and electronic; (d) Autos and diesel; (e) Power; (f) Civil; (g) Textile spinning; (h) Refrigeration and air-conditioning; (i) Chemical; (j) Television; (k) Watch; (l) Metallurgy (until 1975).

Polytechnic Institute, Golimar, Sukkur

Diploma: Radio and television.

College of Technology, Hyderabad

Diploma: (a) Automobile; (b) Civil; (c) Electrical; (d) Mechanical; (e) Tool design and tool-making; (f) Glass and ceramics.

College of Technology, Khairpur

Diploma: (a) Electrical; (b) Mechanical; (c) Civil; (d) Auto and farm; (e) Refrigeration and air-conditioning.

Pakistani/Swedish Institute of Technology, Karachi

Diploma: (a) Clothing; (b) Wood-working; (c) Welding; (d) Electrical; (e) Mechanical.

Jinnah Polytechnic Institute, Karachi

Diploma: (a) Automobile; (b) Mechanical; (c) Civil; (d) Electrical.

Jamia Millia Institute of Technology, Malir City, Karachi

Diploma: (a) Civil; (b) Radio and television.

Saifee Eide Zahabi Institute of Technology, Karachi

Diploma: (a) Electrical; (b) Mechanical; (c) Civil.

Habib College of Technology, Nawabshah

Diploma: (a) Chemical; (b) Electrical.

Polytechnic for Women, Sukkur

Diploma: Garments.

Polytechnic for Women, Karachi

Diploma: (a) Garments; (b) Radio and television; (c) Architecture.

College of Technology, Karachi

B.Tech.: (a) Civil; (b) Mechanical; (c) Electrical.

College of Technology, Hyderabad

Diploma: (a) Civil; (b) Electrical; (c) Mechanical.

College of Technology, Khairpur

Diploma: (a) Civil; (b) Electrical; (c) Mechanical.

Habib College of Technology, Nawabshah

B.Sc.: Industrial technology.
B.Tech. (Hons.): Chemical technology.

North-West Frontier Province

College of Technology, Peshawar

Diploma: (a) Auto; (b) Electrical; (c) Mechanical; (d) Civil; (e) Radio and electrical; (f) Chemical.

Polytechnic Institute, Haripur

Diploma: (a) Electrical; (b) Mechanical; (c) Civil.

Polytechnic Institute, D. I. Khan

Diploma: (a) Electrical; (b) Mechanical; (c) Civil.

Polytechnic Institute for Women, Peshawar

Diploma: (a) Commerce; (b) Architecture; (c) Radio electronics; (d) Garment.

Polytechnic Institute, Mingora

Diploma: (a) Electrical (from 1986); (b) Mechanical (from 1986).

Polytechnic Institute, Nowshera

Diploma: (a) Electrical; (b) Mechanical; (c) Civil.

Case-study for the United Kingdom

James Murray,
Napier College, Edinburgh
and
Wilfred Fishwick,
Department of Engineering Science,
University of Exeter, Exeter

Contents

Engineering and technical education in the United Kingdom—the early years and subsequent developments in Scotland

Introduction

In all recently industrialized, and industrializing, centres the local 'industrial revolution' has been guided and implemented by educated and trained engineers and technicians. In the early stages of industrialization the engineers in this group, and probably most of the technicians, whether of local origin or foreigners, will have had their formation outside the developing country in some already industrialized country. Sending local people abroad to obtain an education and training in engineering is a relatively new practice and although not unknown was quite uncommon when industrialization was accelerating in what are now countries with long-established industries, such as France, the Federal Republic of Germany and the United Kingdom. However, those and other countries did exchange and import information about science and technology, and engineers from one country would work in another. Thus following the successful operation of the steam-powered railway in England hundreds of British railway engineers and technicians went abroad either to build new railways or to advise on their construction. There was also a concurrent export from the United Kingdom of iron and steel components such as rails and engines for these new railways.

The worldwide spread of industries of a new type started with the Industrial Revolution in Western Europe, and the centre of gravity of this revolution was in the United Kingdom, and principally in England and Scotland to begin with. This Industrial Revolution which started in the early years of the eighteenth century is often now called the First Industrial Revolution to distinguish it from the computer-based Second Industrial Revolution which has its origins in the United States of America.

The First Industrial Revolution progressed rapidly during the hundred years from about 1730 to 1830, during which time the United Kingdom became the most industrialized country in the world and produced a vast range of products. Most products were based on the use of iron in new ways and in new artefacts. It has been observed that this advance of technology took place in the complete absence of technological or engineering education. Indeed this seems to be true, for the UK had no advanced engineering schools until about 1850, whilst countries such as France and Hungary, for example, had schools for engineers by the last half of the eighteenth century. Certainly in the United Kingdom there was no state-organized and state-supported system of technological education until late in the nineteenth century. However, education is not always synonymous with the provision of schools, as F. Smith noted in his book *The History of English Elementary Education*. It is true today and was true in Britain during the period of the Industrial Revolution.

In Part A of this case-study Wilfred Fishwick will examine scientific and technological education and training prior to and during the early years of industrialization in Britain. In Part B Dr James Murray will concentrate on the development of engineering and technical education in Scotland. In Scotland, educational practices were somewhat different from those in the rest of the United Kingdom and had different traditions. Today engineering and technical education is little different from that in the rest of the country, so that present practices in Scotland are generally quite descriptive of technological education in other parts of the United Kingdom.

PART A. THE EARLY YEARS

The present system of engineering and technical education in the United Kingdom is almost entirely supported by national and local governments and comprises a large variety of institutions such as universities, polytechnics, engineering colleges, technical colleges and colleges of further education. There are a few private correspondence schools. The present system is relatively new as we shall see and was completely absent during and after the First Industrial Revolution in Britain (used in this chapter as a synonym for the United Kingdom). These new forms of industry did not arise out of nothing by a stroke of magic but were preceded by what has been called the 'scientific prelude to the Industrial Revolution' by writers such as E. A. Musson and E. Robinson. Furthermore, it will be seen that even in the absence of engineering and technical education institutions, there were mechanisms for the diffusion of scientific and technological knowledge amongst interested people. Even today colleges and schools are far from being the only systems by which science and technology are transmitted from one person to another.

The scientific prelude to the modern industrial age

Today, technological education and industry can be considered as partners in some large enterprise dedicated to producing, operating and maintaining goods for the service or use of people. All governments consider it their duty to provide and support all, or the greater part of, the required educational institutions. The modern industrial era, however, largely began, developed and prospered in countries of the British Isles with little or no government assistance over a period of about 150 years. There were many factors that determined why countries such as England and Scotland should be the centre of this explosion in industrial activity and not other countries in Europe. An excellent discussion of these factors are to be found in the book by George S. Emerson on the social history of engineering. Some important factors are: the final disappearance of remnants of the feudal system and so of much monachal power following the civil wars of the seventeenth century; the freedom of the labourer to find work wherever he could; agricultural improvements providing food for a rapidly growing population; the propensity of the upper and middle classes to invest in commerce and industry; a well-established patent system; and the burgeoning of foreign trade and shipping. In addition there was a surprisingly widespread knowledge of contemporary science and technology.

The Industrial Revolution can be said to have begun around the beginning of the seventeenth century, and it is of interest to know how scientific and technological knowledge had been spread so widely in countries of the British Isles before the onset of major industrial advances. The Renaissance of philosophical and scientific inquiry had started in the Italian states around about the beginning of the fourteenth century and had spread across western Europe. This new learning gave rise not only to a flowering of the visual arts and literature but also to many advances in natural philosophy. The natural philosophers began to compare experimental discoveries and the craftsman's knowledge with the ancient theories of Greece. They found these theories unrealistic and set about inventing new ones to explain physical phenomena. Books and pamphlets were written, but there was known to be a considerable correspondence about scientific matters written in the mediaeval Latin that all scholars used. Mathematics was a respected and useful subject and in all of the fifty or so universities in western Europe in 1600 it is believed that mathematics was a subject of study. Be that as it may, there was in the countries of the British Isles in the years following the Renaissance a considerable group of scholars in and out of the universities who were quite up-to-date in their knowledge of contemporary science, and the 'mechanical arts' which is what today would be called technology.

Out of religious argument in the Roman Church there arose Protestant beliefs and churches which had a particular influence in the northern

European states and in England and Scotland. These reformed religions gave increased importance to individuals and their actions, and Calvin emphasized the relationship between those favoured by God, his flock of course, and the useful work accomplished by applying minds and bodies. In Britain Calvinism had its strongest influence in Scotland where the Church under John Knox ordained, as early as 1633, that each Presbyterian Church parish should support a school which would fit young people for a useful life. Of course there were those parishes that did not have such schools, but many did and the spread of an elementary education in reading, writing and arithmetic to some people in all classes meant that the fraction of the population with some education was larger in Scotland than elsewhere. This would turn out to be of some importance when industrialization began, for industrialization and education of the population go together.

In England and Wales the authorities did not provide elementary education for the labouring and the lower artisanal classes until the nineteenth century. However, according to G. M. Trevelyan, as early as 1400 there were between 300 and 400 grammar schools in England and Wales, and this number increased in succeeding centuries. These schools were supported by endowments from individuals and merchant and craft guilds and provided schooling for laymen as well as those destined for service in the Church. Students had to pay small fees but many schools gave free education to the relatively poor from the lower middle classes, rather than from the labouring classes. England, surprisingly, had a rather good secondary system of education before it had a more universal system of primary education. Thus by the beginning of the eighteenth century there were considerable numbers of people who could read and write, who knew some mathematics, and had some knowledge of Latin and history. These people were the source of clerks, merchants, civil servants, architects and designers and scholars. Nobility attended these grammar schools to a certain extent, but in due course some of these schools with special endowments catered primarily for the wealthier classes.

The labouring classes had little chance to attend schools until the Church of England, under pressure from the Non-conformists, founded some hundreds of charity schools between 1700 and 1715 in order to give elementary education and instruction in Church principles. However, there had been, and remained, many private teachers or schools in villages and towns which gave some education for very small fees. These schools were often run by widows or indigent ladies and were called then 'Dame-schools'. There were therefore some opportunities for the labouring classes to gain some sort of education long before state schools existed.

Technological education, as such, did not exist in Britain, or indeed in Europe, until the Industrial Revolution was well under way. However, well before 1700 the works of Copernicus, Descartes, Galileo, Comelius, and many others were well-known in England and Scotland. The brilliant thinker Francis Bacon (1516–1626) expressed his ideas on teaching education and technology in his tract 'The Usefulness of Experimental Natural Philosophy', in 'New Atlantis' and other writings which have been discussed fully by B. Farrington in his works. Sir Francis Bacon was one of a new upper-middle-class 'gentry' whose wealth derived from agriculture, trade, law and administrative posts. Many had the taste and leisure to consider new discoveries and ideas. Bacon was interested in the ways in which science could improve manufacturing and so provide enhanced standards of living. He realized that new educational institutions would be needed in England to provide suitably educated people. His far-sighted proposals included several types of schools, including trade schools, a kind of technological school or college, and a museum and demonstration laboratory, which he called a House of Solomon, which was also to be a research institute staffed by craftsmen and scholars. His bold ideas implied co-operative action to achieve specific ends. They had been foreshadowed by the ideas of Sir Thomas More and others, but even so the contemporary society as a whole was not sympathetic and the ideas were not implemented. Baconian ideas, however, were quite persuasive amongst thinkers and scholars for a long time and spread widely in Britain and its colonies.

Two very important results for science and technology did arise from Baconian philosophy. One was the founding of a new college, in 1679, by Sir Thomas Gresham. Gresham College in the City of London had seven professors, two of whom were both mathematicians and astronomers. The college had both utilitarian and scholarly purposes and had close relations with naval officers, navigators and shipbuilders and thus with instrument- and clock-makers. The other result was the founding of the Royal Society of London in 1662. The Royal Society grew out of a meeting at Gresham College in London (and in Oxford during some years of the Civil War) of Francis Bacon and his followers and friends. The members of this society were principally concerned with the use of new science in manufacture, agriculture and trades, and lists of processes were made which might be useful to study. Natural philosophy was not divided sharply into pure and applied science. Nevertheless, in investigating the mechanical arts, as well as natural phenomena, the members of the society made many new discoveries of a purely scientific nature. Their work was backed up, especially in London, by craftsmen in many trades, and particularly by instrument-makers. Experimental scientific investigation of processes and phenomena was being practised in what might be called 'modern methods'.

Most craftsmen, such as builders, millwrights, mechanics, instrument-makers, shipwrights and so on, learnt their crafts as apprentices to master-workers. Wealth or property went from father to eldest son as a rule, and other sons had to make their own way in the world. If the sons of middle-class or upper-class parents, these sons might be quite well-educated but, having little money, would take up apprenticeships and clerkships. Their education and ambition often helped them to be wealthy men in their turn, but it did mean that other apprentices and craftsmen were encouraged to improve their own education, so that it was by no means true that, in the seventeenth century, for example, craftsmen were illiterate. On the contrary, it was a great era of self-education and it is doubtful if the Industrial Revolution would have taken off so quickly had there not been so many literate and skilled craftsmen in Britain.

Self-education depended on lecturers giving free lectures to interested people, or on lecturers making a living by charging a few pence per lecture or a few shillings for a course of lectures. Arithmetic and mathematics were very popular and well-attended subjects, as were lectures on chemistry and practical arts. Centres for social and philosophical discourse and education from the middle of the seventeenth century onwards were the coffee-houses, especially in London and other important cities. Coffee and chocolate drinks were served in these houses, which acted not only as clubs but also as disseminators of political and business news. Many prominent people gave free lectures; one of them was Dr John Desaguliers who discoursed on 'mechanical and experimental philosophy' at his house in London from about 1713 onwards. He also wrote textbooks on 'experimental (natural) philosophy'. But there were many others, and even professors at Cambridge and Oxford universities found it satisfying to give free lectures to anyone who cared to listen on mathematical and scientific matters. However, the centre of scientific ideas and applications in England remained London.

The word 'engineer' was used only for the military engineer concerned with fortifications and demolitions, and the provision of temporary bridges: that is until the middle of the eighteenth century when John Smeaton first used the title 'civil engineer' to describe himself. Before that time people who did engineering works were called architects, builders, surveyors, 'mechaniks', instrument-makers, millwrights and various other specific names. Smeaton himself began as an instrument-maker and then became a millwright and eventually an engineer. Millwrights were extremely important craftsmen, for there were in existence many thousands of water-driven and wind-driven mills for the grinding of wheat and other cereals. Even in 1086 some 5,000 water-mills were counted in Norman England alone. The millwright built and maintained these mills and inevitably had a knowledge of mechanics, hydraulics and structures, some of it gained during apprenticeship and by experience and some often obtained by attending lectures on mathematical and natural experimental philosophy. As mines for coal and ores became deeper so some millwrights began to construct pumps and drainage schemes for them. Coal-mining was an important activity by 1500, with 500,000 tonnes of coal shipped annually out of ports in north-east England alone. Mining engineers therefore existed, even if they were not called by such a name.

Technological education during the Industrial Revolution

Although water power remained important to the textile industry until the early years of the nineteenth century, and longer for the grinding of corn, the Industrial Revolution was based on the use of steam-powered machines and on the availability of large quantities of cheap iron. At the same time large advances were made in applied chemistry for products used in the textile industry, in explosives, etc. The discovery by Abraham Darby in 1709 that coke could be used to smelt iron ore revolutionized iron-making in due course, as Cort and others developed iron foundry and the use of cast iron. The steam engines of Savery and Newcomen were improved by a large number of workers, including the illustrious James Watt and Matthew Boulton.

However, this is not a history of the Industrial Revolution in Britain, and anyone interested could consult the books in the Bibliography. Here we shall devote ourselves to how scientific and technical knowledge was diffused in Britain, not only to those with adequate schooling but also to those with little or none. The motivation for most self-instruction was that society accepted entrepreneurs, and that opportunities for economic advancement were open to all those who could grasp them. As G. M. Trevelyan has written, 'it was an age of aristocracy and liberty; of the rule of law and the absence of reform; of individual liberty and institutional decay; . . . of creative vigour in all the trades and arts that serve and adorn the life of man'. The feudal system and the peasant society had long disappeared and divisions of class and race hardly existed.

The coffee-houses remained centres of intellectual argument and the meeting places of innumerable philosophic and scientific societies. The Society of Arts—now the Royal Society of Arts—was founded in 1754 in Rawthmell's Coffee House in London. Indeed the coffee-houses were dubbed the 'Penny Universities' and the Royal Society of London often held its meetings in coffee-houses.

Itinerant lecturers on science and technology such as Adam Walker and many others used coffee-houses as lecture rooms if they could not find other premises. Further information on scientific societies in Britain and Europe can be found in the works of Professor W. H. G. Armytage. The word 'Arts' in the Society of Arts meant the mechanical arts, i.e. trade and industrial arts, rather than pictorial arts.

The London Society of Engineers was founded in 1771 and included many well-known engineers, of whom one was John Smeaton. When Smeaton died in 1792 the society changed its name to the Smeatonian Society. Finally, in 1818 the society reformed as the Institution of Civil Engineers, the prototype of the professional engineering institutions of today. Its main purpose was originally the dissemination of engineering information. Other cities had similar societies, the most well-known probably being the Lunar Society in Birmingham. It has amongst its members Joseph Priestley, Erasmus Darwin, James Watt, Matthew Boulton, William Small (mathematician) and many other important industrialists and chemists.

There was the Military Academy at Woolwich, founded in 1741, which devoted itself to teaching military engineering, though presumably some military engineers mixed with their civil contemporaries. It had some professors of mathematics, however, who did publish books on engineering matters. Surprisingly, this school preceded that at Mezières in France, for the French military engineers were at that time the best trained and organized group in Europe.

Outside the coffee-house societies there were numerous lecturers moving around the country or based in large towns earning their living by expounding on arithmetic, chemistry and mechanics to any listener prepared to pay a few pence or a few shillings. This availability of teaching and the eagerness of many working men to learn, and pay for learning, astonished French observers. Whilst a system of primary schools and mechanics institutes organized by the government might have produced better results, the system that existed in the eighteenth century cannot be said to have been a failure. By 1800 there were probably around 4 million industrial workers in Britain, working usually in pretty horrible conditions, but with opportunities to get on in the world. The Charity Schools in England and Wales, Church Schools, and the Parish and Burgh Schools in Scotland ensured that a large number of those workers had some basic, if minimal, education. The writing of elementary manuals on mathematics, science and the arts was almost a major industry, with some hundreds of books available. Some, no doubt, were of poor quality, but they were cheap and available. There were also free reading libraries.

The two universities in England at Oxford and Cambridge had professors of mathematics, and chemistry, and indeed nearly all students had to show some proficiency in mathematics. However, it was the various professors of chemistry, often concentrating in their lectures on the chemical processes in industry, that kept closest to the new industrial society. The Chair of Natural Philosophy at Cambridge was not founded until 1792, and its first holder, Isaac Milner, lectured not only on chemistry, but also discussed steam engines, pumps and other machines. A professor of chemistry called Farish gave lectures from 1796 onwards on the 'Arts and Manufactures, as they relate to Chemistry' and also lectured on machines and mechanics. He drew large audiences at the university. Similarly, the professors of chemistry at Oxford concerned themselves with the arts of manufacture as well as with science of all kinds.

It was in Scotland, as will be seen in Part B of this account, that the connections between science and industry at this time were particularly strong.

As time went on, and the activities of industry multiplied, the need for more formal methods of obtaining technological knowledge became more and more obvious. The centre of gravity of industry and practical science was now outside London, and there were many scientific societies in cities such as Birmingham, Manchester, Derby, Newcastle, Plymouth, Leeds, Sheffield, Bristol and Bath, but there had arisen a group of academies, or 'dissenting academies' as they were called, because they were founded by Non-conformist Church members. These academies taught mathematics and the principles of science and the mechanical arts. The most important was Warrington Academy (1757–86), but there were others in Kendal, Hackney, Manchester, Northampton, Kibworth, York, Salford and so on. Their teachers fostered the growth of mathematical societies, especially in the County of Lancashire. The students who left these academies often helped to start up all kinds of colleges to train workers in clerical and business practices, as well as giving general further education. They tended to come and go, but nevertheless provided a very useful service.

Then there were the famous Mechanics' Institutes of the late eighteenth and early nineteeth centuries. These began because craftsmen or mechanics were eager to be educated further in their crafts. In Glasgow, a professor of natural philosophy at the university, John Anderson, gave a series of public lectures which attracted large audiences, and when he died in 1796 he left a collection of instruments, books and museums plus a detailed plan for a new institution. This was incorporated as the Anderson's Institution in 1796, with Thomas Garnett in the first Chair of Natural Philosophy. Under various heads, including George Birkbeck (1799–1804), the institution survived and became Anderson's University in 1828. The history of this institution is found below in Part. B. What is of interest here is that in 1823, following an argument about the use of a special mechanic's library, a group of mechanics got together and founded the Glasgow Mechanics Institute, a self-governing college. This was the forerunner of hundreds of Mechanics' Institutes in the towns of Britain, and by 1850 there were 610 such institutes with 102,050 members. These were self-governed by their members, who were also the students. In time they became more middle-class than working-class, but while they existed they provided a wonderful service.

The Anderson's Institution was also the model for the Royal Institution in London, founded by Count Rumford in 1800. It was also a model for the School of the Arts in Edinburgh, Greenock and Aberdeen. Dr Birkbeck went to London in 1804. There the founded the London Mechanics Institute and his name is remembered through Birkbeck College, a part of London University.

The Mechanics' Institutes died out in the late nineteenth century, by which time some universities, especially the new ones, were teaching engineering, and technical and vocational instruction was becoming available in technical colleges. The story in England and Wales is a complicated one, and Part B of this account is devoted to developments of technological education in Scotland from the Industrial Revolution to the present day. However, Appendix A.I lists UK university institutions which have engineering courses, and Appendix A.II those polytechnics and central institutions with engineering courses. Polytechnics and central institutions give degrees controlled by the Council for National Academic Awards (CNAA), and there are very many colleges of further education and technical colleges controlled by local governments which also prepare students for CNAA degrees in engineering. These are not listed here.

Bibliography

ARMYTAGE, W. H. G. *The Rise of the Technocrats*. 1965.

——. *A Social History of Engineering*. 4th rev. ed. London, Faber, 1976.

BUCHANAN, R. A. *Technology and Social Progress*. Oxford, Pergamon Press, 1966.

DEANE, P. *The First Industrial Revolution*. 2nd rev. ed. Cambridge, Cambridge University Press, 1980.

EMERSON, George S. *Engineering Education, A Social History*. 1973.

FARRINGTON, B. *Francis Bacon, Philosopher of Industrial Science*. 1951.

——. *Francis Bacon, Pioneer of Planned Science*, 1963.

FLEMING, A. P. N.; BROCKLEHURST, H. J. *A History of Engineering*. 1915.

LANDES, D. S. *Unbound Prometheus: Technological Change and Industrial Development in Western Europe from 1750 to the Present*. Cambridge, Cambridge University Press, 1969.

MUSSON, E. A.; ROBINSON, E. *Science and Technology in the Industrial Revolution*. Manchester University Press, 1969.

TREVELYAN, G. M. *English Social History*. Rev. ed. London, Longman, 1978.

WEST, E. G. *Education and the Industrial Revolution*.

WOOD, A. T. *Industrial Engineering in the Eighteenth Century*.

APPENDIX A.1

UNIVERSITIES IN THE UNITED KINGDOM PREPARING STUDENTS FOR ENGINEERING DEGREES

Aberdeen
Aston
Bath
Belfast (Queen's University)
Birmingham
Bradford
Brunel
Cambridge
City
Dundee
Durham
Edinburgh
Essex
Exeter
Glasgow
Heriot-Watt
Hull
Kent
Lancaster
Leeds
Leicester
Liverpool
London
- Chelsea and King's College
- Imperial College
- Queen Mary College
- University College

Loughborough
Manchester
Manchester Institute of Science and Technology
Newcastle upon Tyne
Nottingham
The Open University
Oxford
Reading
Salford
Sheffield
Southampton
Stirling
Strathclyde
Surrey
Sussex
Wales
- Institute of Science & Technology
- University College of Cardiff
- University College of North Wales
- University College of Swansea

Warwick
York

and also
Cranfield Institute of Technology
with Royal Military College
and National College of
Agricultural Engineering, Silsoe

APPENDIX A.II

POLYTECHNICS AND CENTRAL INSTITUTIONS WITH ENGINEERING COURSES

Brighton
Bristol
Central London
City of Birmingham
City of London
Hatfield
Huddersfield
Kingston upon Thames
Lancashire (Preston)
Lanchester (Coventry)
Leeds
Leicester
Liverpool
Manchester
Middlesex
Newcastle upon Tyne
North London
North Staffordshire (Stoke on Trent)
North-East London
Oxford
Plymouth
Portsmouth
Sheffield City
South Bank (London)
Sunderland
Teesside (Middlesborough)
Trent (Nottingham)
Ulster (Newtownabbey)
Wales (Pontypridd)
Wolverhampton

Bell College of Technology (Hamilton)
Dundee College of Technology
Glasgow College of Technology
Napier College of Commerce and Technology (Edinburgh)
Paisley College of Technology
Robert Gordon's Institute of Technology (Aberdeen)

PART B. TECHNOLOGICAL EDUCATION IN SCOTLAND

Technical education in the universities

The growth of technical education in Scotland was centred on the activities of the four ancient universities: St Andrews, Aberdeen and Glasgow (all founded in the fifteenth century) and Edinburgh (founded a century later). The universities of Scotland had long provided a broadly based education like that of the universities of Continental Europe. This linking of the education system of Scotland to that of the rest of Europe as opposed to that of England was the result of the centuries-old antagonisms between the two countries that were only resolved following the uniting of the two crowns in the early seventeenth century. This separation of the two nations meant that Scottish scholars on completion of their undergraduate education in Scotland frequently embarked on a tour of study in the great universities of Europe, particularly those of Flanders. Calvinism linked Scotland to Holland and Switzerland. This contact of Scottish scholars and the mainstream of European scientific thought had a considerable influence on the subsequent development of the Scottish universities.

In addition, the universities of Scotland have traditionally admitted their students from a wide spectrum of the population and, with much of the wealth of the country after the Industrial Revolution being generated by its manufacturing industries, large numbers of high-quality students have always been attracted to courses in the applied sciences and engineering and naval architecture. The majority of students, however, studied for the ministry or to become schoolteachers, and the curriculum was designed primarily for their educational needs. In addition, standards were difficult to maintain as the Scottish tradition had been to admit all entrants without reference to their previous scholastic career. The Scottish education system had many advantages, but it also suffered from two major drawbacks that were to impede the development of scientific and engineering education: first, there were no honours degrees, the system being based on the broadly based ordinary degree; and second, the absence of a mathematical tradition in Scottish schools. The lack of honours degrees meant that there was little specialism and the lack of mathematics seriously affected the ability of students to understand the concepts of physics and engineering. The only mathematician of any status produced in Scotland, John Napier of Merchiston, famed for his work on logarithms and spherical trigonometry, had obviously gained his knowledge of mathematics in the universities of Flanders and Germany, and he founded no Scottish school of mathematics.

However, the universities of Scotland had long included classes in philosophy in their curricula and as early as 1727 a statute of the

University of Glasgow decreed that classes in experimental philosophy would be open both to students of the university and to members of the general public. John Anderson, professor of natural philosophy (physics) (1757–96), encouraged members of the public to attend his classes and he supplemented his lectures in pure physics with a class in experimental physics which he held in the evenings for extramural students.

Anderson died in January 1796 and in his will left instructions regarding the founding of a technological university, although he did not leave sufficient funds to endow its foundation. However, soon afterwards his ideas became reality with the establishment of Anderson's Institution in which the tradition of providing extramural classes was not only continued but was considerably expanded, and this led to the development in parallel with that of undergraduate education of an equally important system of non-graduate technical education.

The early university courses in engineering in Scotland were a logical development of the courses in natural philosophy that played an important role in the development of the universities in Scotland. The studies of natural philosophy and medicine and chemistry were to be the basis of studies of the applied sciences in the universities of Scotland. It is also significant that there were important differences in the treatment of the sciences in the various universities of Scotland and those differences were still apparent in the philosophy and format of the courses over a century later. The University of Edinburgh soon established a reputation for the excellence of its pure research, whilst the University of Glasgow and Anderson's University, the world's first technological university and the forerunner of the University of Strathclyde, were closely involved in the application of their discoveries to practical situations.

These differences were perhaps best exemplified by the contrast in careers of their two internationally famous professors of natural philosophy. Clerk-Maxwell at Edinburgh pursued theoretical studies of electromagnetism later to be developed and applied by others. William Thomson, later to become Lord Kelvin, at Glasgow gained his fame first through his studies which led to the formulation of the fundamental laws of thermodynamics and second by his work on underwater telephone cables, the latter gaining him a knighthood. Kelvin was also the holder of a number of important patents which provided him with a large annual income.

In earlier times Joseph Black (1728–99) had started the process of co-operative education between academics and industry. The important discoveries by Black of latent and specific heat led to him providing assistance to industrialists to improve the quality of their products using his newly discovered knowledge.

During the nineteenth century, government commissioners were appointed to investigate the education system and the universities of Scotland. They were responsible for sweeping changes which were introduced in stages, for example the introduction of Regius Chairs in a number of new disciplines, including engineering at Glasgow in 1840. This was the first chair of engineering in any university in the world (by 1914 Glasgow had three chairs of engineering and one each in naval architecture and mining).

Some of the recommendations of the commissioners were strongly resisted by many of the existing professors at the universities, some believing that such subjects as engineering had no place in a university. The replacement of the Scottish general degree with honours by honours degrees resembling the English pattern was also resisted, as it was thought that this would destroy the traditional concepts of a broadly based education. The commissioners were also responsible for the raising of the school leaving age and the introduction of minimum entry standards for university entrance.

The importance of engineering at Glasgow was recognized by the appointment of Lewis Gordon as the first holder of the Chair of Engineering. Gordon was a prominent civil engineer who, after his appointment to the university, continued his work as a practising civil engineer. He was responsible for planning and implementing the supply of Loch Katrine water from the Scottish highlands to Glasgow. He was also an entrepreneur and founded a cable-making company in England which produced the first transatlantic telephone cables. This recognition of the importance of engineering was later consolidated by the appointment of William McQuorn Rankine as the second holder of the chair. Rankine became deeply involved with the activities of the local industries: heavy engineering and shipbuilding. Rankine developed an international reputation for his work on thermodynamics and heat engines and he soon became involved in explaining the relevance of his theoretical studies to the improvement of the design of steam engines for the propulsion of the large ocean liners being built on the Clyde. Rankine also became the founder of the Institute of Engineers and Shipbuilders, demonstrating the high esteem in which he was held by industrialists.

The University of Edinburgh appointed its first professor of engineering in 1868. Fleeming Jenkin, who was appointed to the post, had a reputation in a number of disciplines including science, literature and drama, as well as engineering. Jenkin, however, did not altogether neglect the applications of engineering science to practical solutions. He did important work on ocean telegraphs and assisted in the important development of a system of engineering drawing by which three-dimensional objects can be represented by orthographic projection in two dimensions.

The international reputation of engineering at the University of Glasgow was further enhanced by the work of Professor Alexander Barr. Barr was eminent in the field of applied optics and his studies are still commemorated in the name of the firm Barr & Stroud, famous for the production of rangefinders and other optical devices. It is not without significance that, when the Japanese Government decided to strengthen their industrial base, Glasgow was among the chosen centres and soon rangefinders based on the work of Barr were installed in the warships of the Imperial Navy.

What should have been a significant development in the growth of engineering and scientific education in Scotland was the foundation of a new college in Dundee. This college was the result of a large bequest in 1880 by one of the Baxter family, prominent Dundee industrialists, for a college to support the needs of local industry (chiefly textiles and engineering) in research and technological training. Unfortunately the new college was not allowed to develop into a Civic University on the English model but was linked to the University of St Andrews on the opposite shores of the Firth of Tay. Soon the business community of Dundee became aware that their hopes for the college were not being realized and in the linking with St Andrews their new college was being stifled by the lack of any scientific or technological tradition at Scotland's oldest university.

This perhaps was true of all the universities of Scotland in that the new disciplines of science and engineering were dominated by the classical traditions of the universities. In England, which in 1801 had only the Universities of Oxford and Cambridge and by 1880 still had only four, there had been a tremendous growth in the availability of university education by the establishment of the great Civic Universities. These new universities, unfettered by tradition, had an atmosphere in which science and engineering could flourish.

The Scottish universities, which had enjoyed the advantages of being the first to introduce engineering, had seen their relative position decline, and there was also a switch away from industry as the research carried out in the universities became less applied.

In 1901 another system of engineering education became available to students in Scotland. Paisley College began to offer the external degrees of the University of London, a route used by many Scots unable to attend full-time courses at the Scottish universities.

The development of technological education in the universities of Scotland was closely linked to its indigenous industries of shipbuilding and heavy engineering. These industries were not traditional employers of graduates and in these industries the rate of technological innovation was slow. Therefore a large proportion of Scottish graduates in engineering have sought employment outside Scotland, either elsewhere in the UK or in the Commonwealth. The Scottish engineering industry suffered until the late 1930s through its inability to diversify and develop science-based and precision industries; for example, although there was some work done on airship construction and the development of an autogiro there was no aircraft industry and early forays into automobile production soon foundered because of the distance from the principal markets and lack of component suppliers. The chief exceptions were companies employing the latest production engineering methods such as Albion Motors, now part of British Leyland, and Singer, the American sewing-machine manufacturers, the latter establishing their first and largest European plant near Glasgow in a purpose-built factory (now closed).

It was not until the late 1930s and the impending World War with the establishment of industrial estates and the introduction to Scotland of firms like Rolls-Royce and others involved in electronics that the Scottish universities and their students had access to the essential infrastructure of high-technology firms who appreciated the value of employing significant numbers of graduates.

The Scottish universitities tended to stagnate, and examination of the curriculum of the courses in engineering showed that there was little change in subjects taught over two decades, and that there was an emphasis on the 'engineering science' approach to the teaching of engineering, at the expense of applications and design.

However, the system was not moribund. In 1920 the University of Glasgow modified its engineering undergraduate courses to the sandwich mode. As the Scottish undergraduate honours degrees were of four years' duration this meant that their students had three periods of six-month industrial placement in which they could relate their academic studies to practical situations. This innovation in co-operative education is generally accepted as being the first such course in the UK. This mode of education continued at the University of Glasgow and the Royal Technical College until the 1960s when it was resolved that more time was required for academic studies in view of the rapid changes in technology, particularly in electronics and control engineering. The Scottish universities also did not accept the view held by many engineering educationists that design could not be taught and had no place in the academic curriculum. In 1966 the University of Strathclyde established a Chair of Production Engineering; in 1969 it was endowed by Rolls-Royce, when it became known as the Rolls-Royce Chair. A Chair of Engineering Machine Design was established in 1893 in the West of Scotland Technical College, this being encapsulated in the Mechanical Chair in 1904 and merged as the Engineering Design Chair in 1971.

In addition, in an act of close co-operation between the University of Glasgow and the Royal Technical College (RTC), Glasgow, certain students of engineering and mining enrolled at the latter were able to read for degrees of the former. In addition the RTC ran courses leading to the award of diploma and associateship awards of degree and honours degree equivalent. In the east, relationships between Heriot-Watt College and the University of Edinburgh did not develop along such harmonious lines and the Heriot-Watt College offered only diplomas and associateships of degree and honours degree equivalence. The diploma and associateship courses of the two colleges proved very attractive to a large number of very able students who became successful engineers and managers after graduating but who were unable to gain entry to the universities, which insisted at that time on entrants having a pass in a foreign language in their entry qualifications.

The years subsequent to 1945 saw the greatest rate of development in engineering education in Scotland, as in the rest of the United Kingdom. The Royal Technical College became the Royal College of Science and Technology in 1956 and merged with the Central College of Commerce and received its university charter in 1964. The Robbin's Report published in 1963 had proposed that five Special Institutions for Scientific and Technological Education and Research (SISTERS) should be designated. The University of Strathclyde was to be one of the institutes; however, this imaginative proposal was not implemented, though three universities—Strathclyde, Imperial College and UMIST—were the recipients of special grants in succeeding years. The Heriot-Watt College became Heriot-Watt University in 1966, and in Dundee, Queen's College became the University of Dundee in 1967, finally realizing the aspirations of its original benefactor.

The new technological University of Strathclyde had over 90 per cent of its students on courses in science and technology, but the university soon expanded and diversified, adding a wide range of disciplines in the arts and social sciences. Engineering thus declined in its relevant standing in the university. Heriot-Watt University still had a wide range of courses below degree level and these were transferred to the newly built Napier College, with significant implications for the development of higher education in Edinburgh.

Several factors have had a significant effect on the development of higher technological education in Scotland in the period since the 1950s. The continual and accelerating decline of the traditional engineering industries of the Central Belt of Scotland, the decline in coal-mining and the end of shale-mining was balanced by efforts by government departments to attract multinational firms to locate in Scotland, e.g. Burroughs, Caterpillar, Euclid, Hoover and Cummins in the west of Scotland; NCR, Veedor Root, Timex in Dundee and in Ferranti Ltd, Edinburgh, which was a seed bed for the growth of the electronics industry in the east of Scotland. In addition, the National Engineering Laboratory (NEL) was based at East Kilbride near Glasgow. Government regional policy provided funds which resulted in the motor-car industry briefly returning to Scotland through Rootes at Linwood and BMC at Bathgate.

The supply of graduate engineers from the Scottish universities backed by suitable training facilities for technicians and the interface in the Scottish colleges plus an adequate supply of skilled labour from the declining industries and government incentives made Scotland an attractive place for new plants to be located and this was given a further boost by the UK joining the EEC. This first phase of location in Scotland was followed by an all-out drive to attract multinational firms in the electronics industry to site their new factories in Scotland and this meant that firms such as IBM, Hewlett-Packard, National Semi-Conductor and Mitsubishi are now fully established in the Central Belt of Scotland.

The discovery of oil in the northern North Sea in the 1970s added a further range of new technologies to the industries of Scotland as firms became involved in the exploration for oil, and later the production of oil from offshore fields. Yards for the building of offshore structures were established around the coast of Scotland and the depth of water in the North Sea and the general environmental conditions meant that the metal and concrete structures being designed for use in these conditions were at the forefront of the new technology.

This range of new industries provided the high-technology base previously lacking in Scotland and the opportunities afforded by the new industries have had a significant effect on the curriculum of Scottish engineering courses and also on the industrial involvement of Scottish educational establishments. In addition, the advent of the CNAA (Council for National Academic Awards) offered the opportunity to the major Scottish colleges (Paisley College of Technology, Dundee College of Technology, Robert Gordon's Institute of Technology, Aberdeen, Glasgow College of Technology, and Napier College, Edinburgh) to offer degree courses in engineering under the auspices of the CNAA.

It is interesting to note that the majority of CNAA degree courses are either of the sandwich pattern or incorporate a six-month period of industrial training. Such courses are normally of four years (degree) or four and a half years (honours degree) duration whilst the honours courses of the universities last four years. The extra year spent at universities and colleges by Scottish undergraduates compared with students in England reflects the fact that Scottish pupils complete their school education a year earlier than pupils in England.

The Scottish universities and colleges devised a wide range of new courses, both graduate and undergraduate, to meet the needs of industry and which incorporated the skills of the new technologies. Undergraduate courses offered include electronic engineering, energy engineering, industrial/production engineering, industrial design (technology), off-shore engineering, transportation engineering, together with courses combining the skills of engineering with management. Postgraduate courses include computer-aided engineering, digital systems, pressure vessel design and oil drilling technology.

The Scottish educational institutions also eagerly seized the opportunity of developing links with industry and much of the research being carried out is of an applied nature. Perhaps the most interesting development has been the establishment of the Wolfson Microelectronics Institute as a centre for product innovation and exploration in the field of microelectronics research. This unit, strongly coupled to the University of Edinburgh, was established to improve the training of engineering in microelectronics through exposure to industrial developments and practices, to introduce new microelectronic technology into industry and to exploit commercially the research ideas conceived in the university.

The institute has two major technological thrusts to its activities: integrated-circuit component design and microelectronics product development. The unit has established a marketing organization, Inmap Ltd, which has several spin-off companies and has a sales revenue of over a million pounds per annum.

In 1969 the University of Strathclyde was one of six universities chosen to house a Ministry of Technology-funded industrially based unit. These units were designed to act as powerhouses to surrounding industry, to take on research contracts and to stimulate research in industry. The Strathclyde-based Centre for Industrial Innovation was a leading example of these university centres.

Just recently the University of Strathclyde has established a large unit to ensure that engineering undergraduates at the university receive an appropriate training in basic engineering skills.

The Heriot-Watt University and Paisley College of Technology together were chosen by the Ministry of Technology to provide Low-Cost Automation Centres based on the model of TNO/University of Delft in the Netherlands. These centres were designed to introduce automation into companies in both the east and west of Scotland. The Heriot-Watt centre assisted development of Unitube, a centre designed to co-ordinate the consultancy links between academics and industry. Similarly, the centre at Paisley led to the development of the Microelectronic Development Centre which provides a service in the assessment of new microelectronic hardware and software for industry and education.

Several of the Scottish universities and colleges have also developed teaching companies in which industrial companies and educational institutions combine in the solution of industrial problems using new graduates. This interesting and most successful scheme is funded jointly by the Science and Engineering Research Council (SERC) and industry. Similarly, Napier College is in the forefront of FMS research in the UK and this has been developed in conjunction with a number of industrial companies.

In 1985 the importance of high-technology industry to the future of Scotland was recognized by the Scottish Education Department who provided funds to assist colleges to switch to technology. This 'switch funding' provides for the provision of increased numbers of academic and support staff and equipment to attract additional students into branches of engineering associated with microelectronics. The University Grants Committee has followed this with a similar scheme in certain universities.

Implementation of the requirements of the Engineering Council for corporate membership of a chartered engineering institution have caused both universities and colleges to critically appraise the curriculum of their engineering courses.

The requirement of the Engineering Council is that courses should incorporate an appreciation of engineering skills and a knowledge of the properties of engineering materials (EA1). Also throughout the course there should be an emphasis on design and engineering applications (EA2). In addition undergraduates should have a knowledge of economics, marketing, the organization of industry and the role of the engineer in society.

The changes required to the undergraduate curriculum have been further complicated by a realization that undergraduates should gain practical experience of computer-aided engineering.

Incorporating all these new subjects involved universities and colleges in considerable expenditure for the provision of new facilities and the training of both academic and support staff.

The majority of college courses have been revised and are now designated B.Eng. or B.Eng. (Hons.), the title introduced by the Engineering Council and CNAA to show that they comply with their new requirements. It is still unclear (1985) as to whether the universities will exchange the award of B.Sc. for that of B.Eng. or whether they will introduce sandwich training into their courses.

Technical education in the non-university sector

Technical education in Scotland outside the universities began with the foundation of Mechanics' Institutes and also with the popular lectures given by professors at the universities.

Part-time education in Scotland has a long history. As early as 1709 the Scottish branch of the Society for Promoting Christian Knowledge made it part of the job of the teachers they were training to be sent to the Scottish Highlands to conduct part-time evening classes in the '3 Rs' (Reading, Writing and Arithmetic) in addition to their other duties of conducting day schools for children.

In the village of Methven, near Perth, between the years 1807 and 1817, Thomas Dick, Master of the Secession Church School, established the first Mechanics' Institute in Scotland.

Mention has already been made of the statute of the University of Glasgow in 1727 allowing the opening of its classes in experimental philosophy to members of the general public. In 1880 Professor Birkbeck who succeeded Professor Garnet, the first professor to be appointed at Anderson's Institute, presented to the Governors of Anderson's Institute his proposals for courses of instruction for tradesmen to be held on

Saturday evenings. Birkbeck also indicated that he would teach the class without payment and that rent would be paid for the use of a room.

The Governors were very sceptical as to the value of such classes and they suggested 'that if invited, the mechanics would not come, that if they did come, they would not listen and if they did listen, they would not understand'. Birkbeck had already noticed that the workmen who had been responsible for the construction of machines for his workshops had a considerable interest in understanding the principles of operation of the apparatus and was confident of the success of his proposals.

To the delight of Birkbeck, 75 attended on the first Saturday evening, 200 the second, 300 the third and over 500 on the fourth Saturday evening. Such classes were continued by Birkbeck until his resignation in 1904. The syllabus of the work for the first mechanics class was prepared after a detailed analysis of the work done by the mechanics in their daily jobs as blacksmiths, glass-blowers, carpenters, turners, instrument-makers, tin-plate makers, and so on.

In 1823 the Mechanics' Institute in Glasgow separated from Anderson's Institute and gained its own staff and premises. The Mechanics' Institute rejoined Anderson's College in 1886 and became the Glasgow and West of Scotland Technical College in 1887.

Similar Mechanics' Institutes were formed throughout Scotland, initially in the major industrial centres. In the East of Scotland, Leonard Horner instituted a School of Arts with a strong vocational basis. The first prospectus for the School of Arts stated:

> This association has been formed for the purpose of enabling industrious tradesmen to become acquainted with such of the principles of mechanics, chemistry and other branches of science as are of practical application in their several trades, that they may possess a more thorough knowledge of their business and acquire a greater degree of skill in the practice of it, and be led to improvement with a greater degree of success.

By 1824 there were 340 students in attendance and the sum of £500 was available from surplus fees. It was decided to spend this money on a memorial to James Watt in the form of a building, there being, by 1847, nearly 700 students in attendance. In 1851 a new building was erected and the new institution was named 'The Watt Institution and School of Arts'.

In Dundee 'The Watt Institution for the instruction of young tradesmen in the useful branches of the arts and sciences' started its first classes in 1825 and although it closed in 1868 it was succeeded in 1888 by the Dundee Technical Institute.

By 1850 there were approximately thirty Mechanics' Institutes in Scotland and although a number of the institutes had a brief life the majority survived.

In 1868, under the auspices of the 'Society of Arts', delegates from the Mechanics' Institutes and Schools of Arts attended a Conference on

Technical Education. This conference indicated the growing importance of the institutes and reports of its proceedings offer some interesting insights into the work. During the conference a motion was passed advocating a reorganization of the Mechanics' Institutes so that more emphasis could be placed on scientific teaching. The view was also expressed that the principles of science should form an important element in the tuition of all classes of society. The speakers at the conference included Lyon Playfair, Professor of Chemistry at Edinburgh, a significant figure in the development of scientific and technical education in the UK. Playfair represented the universities of St Andrews and Edinburgh as the Member of Parliament for seven years (until the 1950s the universities in Scotland elected members to the UK Parliament). Playfair had been educated in Scotland but he had also studied under Liebig, the famous German chemist, and was thus able to compare and contrast the systems in the two countries. He was also a Special Commissioner at the Great Exhibition of 1851, where he recognized that state intervention was necessary if technical education were to be properly established.

The Mechanics' Institutes in Scotland owed their continued existence to the fees received from those enrolling for classes and from occasional donations from people who recognized the potential of the institutes. After 1853 some limited state aid was made available, and in 1885 the separation of the Scottish Education Department took place with the establishment of the Scottish Office. The Science and Art Department, founded in 1853 after the Great Exhibition, based at South Kensington, London, remained responsible for many aspects of scientific and technical education until 1867 when the Scottish Education Department, as it was then known, assumed responsibility for affairs in Scotland.

Playfair campaigned vigorously for the establishment of a system of technical education in Britain that would be modelled on those of France and Germany, and which would be supported by significant state financial assistance. Playfair attributed the superiority of some of the industries of Britain's competitors to a number of factors, including Britain's lack of technical schools, the failure of its universities to concentrate on scientific and technical education and the low social standing of engineers and technologists.

This concern for the state of British industry and technical education forced the government to establish in 1881 a Royal Commission on Technical Education (The Samuelson Commission).

The report of the commission recommended that technical subjects should be taught in both primary and secondary schools and in teacher-training colleges, that a unified system of primary and secondary education should be established and that special institutions of high standing should be set up to deal with scientific training.

The Samuelson Commission recommendations led to The Technical Schools Scotland Act which allowed local authorities to raise revenue to provide support for the technical schools, and in 1901 guidelines issued by the Scottish Education Department (SED) allowed the establishment of 'Central Institutions' (CIs) modelled on those set up by the City and Guilds of London Institute. The new Central Institutions would allow the development and improvement of technical education to take place. The work to be carried out in the new institutions would be of the highest standard and would be strictly vocational. They were seen by the SED as potential 'technical universities' and also as the apex of a system of colleges of technical education. The SED also indicated that the management of the CIs should be devolved to governors representative of local interests who would be given considerable freedom in the operation of the institutes.

In Glasgow, Edinburgh, Dundee and Aberdeen the Mechanics' Institutes had gradually evolved into technical colleges and in the high schools evening classes were being offered in technical subjects, so there arose a need for a rationalization and standardization of the work done in the evening classes with that done in the elementary classes in the technical colleges. This need for standardization was intensified as more and more students wished to transfer to the classes at the technical colleges. Schemes of transfer and exemption for students were therefore formalized and implemented.

In the west of Scotland, meetings between representatives of the Scottish Education Department, the Glasgow and West of Scotland Technical College and the school boards of Glasgow and Govan led to agreement that the Glasgow and West of Scotland College would cease to teach the elementary stages of their courses which would be taught in the evening centres, and that a scheme would be administered by a Joint Committee on the Organization of Science Classes, and also that the syllabuses to be followed and example papers (term and final) would be prepared by the staff of the Glasgow and West of Scotland Technical College in consultation with the staff of the evening centres.

The First World War led to the cancellation of a most interesting forward-looking proposal by the firm of Albion Motors Ltd, of Scotstoun, Glasgow. They had proposed that their apprentices should be sent to day-release classes for three years. All the arrangements had been made for sixty apprentices when the advent of the war and the taking over of the works for the war effort caused the cancellation of this most imaginative proposal.

The war also brought home to government and industry the continuing need to improve the standards of technical education throughout the UK, and shortly after the end of hostilities proposals were being made for the appointment of a committee chaired by the President of the Institution of Mechanical Engineers and representatives of the Department of Education to report on the education and training of young people in industry. The committee recommended a National Certificate

Scheme throughout the UK. This scheme would have both Ordinary and Higher National Certificates and would replace the proliferation of Junior and Senior certificates issued by the technical colleges. Each college would submit its scheme of work and final examination papers for approval and also student scripts would be submitted for assessment. The first committee to be established was in mechanical engineering and consisted of representatives of the SED and the Institution of Mechanical Engineers (I.Mech.E.), and the National Certificate scheme was introduced in England and Wales in 1921.

There was a reluctance in Scotland to accept the new scheme as it was felt that its standards were below those of the existing college certificates. However, the scheme was adopted to run in parallel with the existing schemes between 1922 and 1929. Nevertheless, it was not until the middle of the 1930s that the number of National Certificates equalled that of the college certificates.

The most significant difference between the two sets of awards was that students sitting the examinations of the National Certificates had to pass all the examinations of each year of the course at one sitting, whereas with the college certificates students were credited with passes in individual subjects and were awarded a certificate when all the relevant subjects had been passed. The new National Certificates had the additional advantage of national and international recognition since they bore the seals of the I.Mech.E. and SED and were accepted by the appropriate professional institutions as satisfying their membership requirements. The introduction of the first scheme in mechanical engineering was quickly followed by similar schemes in chemistry and naval architecture, and in 1928 and 1931 schemes in electrical engineering and building respectively.

Gaining membership of the appropriate engineering or scientific institution remained the chief motivating factor for thousands of young men and the occasional young women who studied by the evening-class route, three nights per week, to improve their future career prospects. The facilities for the Ordinary National Certificates were made available in a large number of evening high schools but the facilities for the Higher National Certificates were concentrated in the Central Institutions and in addition to the time spent in studying students had to spend considerable time in travelling.

The years subsequent to 1948 saw the influx into the west of Scotland and Dundee of a number of American firms based on precision engineering and large batch production techniques. This caused a need to provide an appropriate education for the technicians in such industries, and the subject 'workshop processes' was introduced into the Ordinary National Certificate course in mechanical engineering. A Higher National Certificate course in production engineering was devised and it was expected that personnel would attend the new course. There was a reluctance to accept it. First, senior management in the traditional industries preferred their employees to follow a course leading to membership of the I.Mech.E., and completion of the HNC in production engineering did not satisfy the requirements of the I.Mech.E. and only gave partial exemption from the educational requirements of the Institute of Production Engineers (I.Prod.E.). In addition, the newer I.Prod.E. did not have the same studies as the older I.Mech.E. so students working in the new industries still continued to study the traditional subjects of thermodynamics and fluid mechanics which were largely irrelevant to their industrial requirements. Later, as the Institution of Mechanical Engineers changed their educational requirements towards HNDs and degrees only and the HNC courses failed to satisfy I.Mech.E. exemption requirements, production engineering as a discipline became firmly established and accepted by industry.

In 1943 plans were being made for the development and expansion of part-time education after the war and it was recommended that apprentices should be allowed to attend classes not less than one day per week in specially built colleges, and that the colleges should be staffed with teachers possessing both industrial and educational qualifications. After the war there was an expansion in the subjects being offered and joint certificate courses were introduced in metallurgy, applied physics and for the expansion of the shipping industry a two-year Ordinary National Diploma in marine engineering. The latter was recognized by the Board of Trade and after a further year at sea the student would return to complete the examinations for the Certificate of Competency as a marine engineer.

The immediate post-war years saw a rapid increase in demand for full-time education at the Central Institutions, with a consequent increase in their use of facilities and increased work-load for their staff. Steps were therefore taken to devolve much of their part-time teaching to the new colleges, which had been built in anticipation of these demands, and to the older established Paisley College. These colleges were equipped to allow the student to complete his studies for the HNC and much of the lower-level courses were completed in local centres.

Those developments of a Scottish system of further education for technicians were not matched by any corresponding development in the education of craftsmen, and the schemes of the City and Guilds of London Institute were adopted by the Scottish colleges later as the City and Guilds developed technician courses. There was thus a significant overlapping of courses.

In 1961 the need for an engineering qualification that would satisfy the needs of industry for higher technicians who had received a thorough training in engineering fundamentals was coupled with studies directed toward engineering applications. The result was the introduction by Paisley College of a three-year sandwich course in mechanical and production engineering. These courses were adopted several years later by the CIs in Dundee and Aberdeen and also by the two new major local authority colleges, Glasgow College of Technology and Napier College in Edinburgh. These new courses were quickly followed by courses in electrical engineering and eventually by courses in a wide range of disciplines, e.g. civil engineering, building, chemistry, physics, biology and catering.

Initially, the engineering HNDs provided full exemption for the educational requirements of the relevant professional institutions. Later, students had to satisfy the additional hurdle of the academic test of the CEI (two technical papers) and the Engineering Society. Later still, when the HNDs were not acceptable for corporate membership of the engineering institutions, their format was changed to provide a greater emphasis on student-centred learning and design studies, and they continue to satisfy the industrial requirement for higher technicians.

The changing nature of industry, with the expansion of such sectors as the health service, led to a rapid expansion in the range of disciplines being offered and by 1963 there grew a desire to have a truly National Certificate with common papers being offered in all Scottish colleges. This led to the amalgamation of the previous Regional Committees for Technical Education into a new body to be known as SANCAD (Scottish Association for National Certificates and Diplomas). SANCAD stayed in existence from 1963 until 1973 and the changing pattern of technical education can be illustrated by the changing pattern of attendance over the decade, as shown in Table 1.

By 1973 there were 34 local authority colleges offering SANCAD courses and of these 33 were purpose-built colleges, with 25 of the colleges having been built in the previous twenty years.

Further reorganization of the administration of technical education followed the recommendation of the Hudson Committee that was set up to 'consider the general pattern of courses for technicians (and equivalent non-technical grades) in Scotland'.

The Hudson Committee followed a similar committee in England and Wales, the Haslegrove Committee. The committees recommended the rationalization of technician courses: courses for technicians should take

TABLE 1. Enrolments in technical education (Scotland), 1963 and 1973

	1963	Percentage	1973	Percentage
Evening	7 888	49	1 921	13.4
Day release	7 682	48	8 288	57.7
Block release	165	1	664	4.6
Full time	304	1.9	3 490	24.3
TOTAL	16 039		14 363	

account of the work of relevant training boards and entrance qualifications should be more flexible. What was more important was that technician courses should be devised to satisfy the educational and industrial objectives of the student and not be, as formerly, courses which would satisfy the educational objectives of the professional institutions which had now moved towards all-graduate entry. This meant that courses would be devised by the new body established to replace SANCAD, known as SCOTEC or Scottish Technical Education Council, and that the four joint committee courses would be phased out and replaced by SCOTEC Courses. The title of the new body which incorporated the term 'technical' and not 'technician' reflected the range and level of the new courses to be offered. SCOTEC courses were known as Higher Certificate and Higher Diploma, the 'National' being dropped.

SCOTEC courses also reverted to the principle that subjects successfully studied should be recorded on the appropriate certificate or diploma, the requirement that the student should pass all subjects of a stage at one sitting being removed.

The new SCOTEC courses were devised by educationists with the assistance of relevant industrialists and the new courses showed a considerable shift in emphasis towards student-centred learning methods, including project and case-study work.

In 1984 two further developments took place in the consolidation of technical education in Scotland with the merging of the bodies responsible for technical and commercial education in Scotland under the title of SCOTVEC (Scottish Vocational Education Council), and the introduction by the SED of the 16–18 Action Plan, by which the former certificate courses of SCOTVEC would be replaced by a system of study based on sixty-hour modules over the whole span of commercial and technical education and where the modules would be offered in both schools and colleges, thus bringing about a greater emphasis on vocational education in schools in Scotland.

APPENDIX B.I

DATES OF FOUNDATION OF THE SCOTTISH UNIVERSITIES

St Andrews	1412	Strathclyde	1964
Glasgow	1451	Heriot-Watt	1966
Aberdeen	1495	Stirling	1966
Edinburgh	1583	Dundee	1967

Note. Following the implementation of the proposals of the Robbin's Report the number of universities in Scotland was doubled.

The universities of St Andrews and Stirling do not offer undergraduate courses in engineering. The University of St Andrews previously offered degrees in engineering at University College Dundee, but following the establishment of the University of Dundee all teaching is now carried out by the lecturer. The University of Stirling has a course in computing science and engineering. Students in these classes undertake some of the laboratory work at the Falkirk College of Technology.

APPENDIX B.II

THE DEVELOPMENT OF THE MAJOR SCOTTISH INSTITUTIONS

Anderson's Institution	1796	Royal Technical College	1912
Anderson's University	1828	Royal College of Science and Technology	1956
Anderson's College	1877	University of Strathclyde	1964
Glasgow and West of Scotland Technical College	1887		

Edinburgh College of Arts	1826
Watt Institution and School of Arts	1856
Heriot-Watt College	1885
Heriot-Watt University	1966

Aberdeen Mechanics Institute	1824
Robert Gordon's College and Gray's School of Art	1855
Robert Gordon's College	1881
Robert Gordon's Technical College	1909
Robert Gordon's Institute of Technology	1965

Dundee Watt Institution	1820
Dundee Technical Institute	1888
Dundee Technical College and School of Art	1909
Dundee Technical College	1935
Dundee College of Technology	1965

Paisley Technical College and School of Art	1897
Paisley Technical College	1950
Paisley College of Technology	1963

Glasgow College of Technology	1971

Napier Technical College	1964
Napier College of Science and Technology	1967
Napier College of Commerce and Technology	1974

APPENDIX B.III

COURSES IN ENGINEERING AT THE SCOTTISH UNIVERSITIES AND COLLEGES

	Aero	Chemical	Civil	Electrical and electronic	General	Ind./Production	Metallurgy	Naval architecture	Offshore	Computing
University										
Aberdeen			x	x	x					
Dundee			x	x						
Edinburgh		x	x	x	x					
Glasgow	x		x	x				x		
Heriot-Watt		x	x	x					x	x
Stirling										x
Strathclyde		x	x	x		x	x			
College										
Dundee College of Technology			x	x						
Napier			x							
Paisley			x	x	x					
Robert Gordon's										

Other institutions

Dundee University: Mechanical engineering and electronics; Electronics and microcomputer engineering.

Heriot-Watt: Civil and gas engineering.

Napier: Energy engineering; Technology with industrial studies; Transportation engineering.

Strathclyde: Electronics and microprocessor engineering; Manufacturing science and engineering; Production engineering and management.

Note. Glasgow College of Technology at present only offers part-time degree courses in Engineering.

APPENDIX B.IV

ENGINEERING EDUCATION IN SCOTTISH SECONDARY SCHOOLS

Study of aspects of engineering have traditionally found a place in a number of Scottish schools. However, a number of factors have inhibited the development of engineering education, including the lack of qualified teachers of the subject and the lack of suitable laboratory facilities. The chief obstacles have been the attitude of the universities who refused to accept higher engineering as an entrance qualification. When engineering was replaced by engineering science the subject was seen as a poor relation of physics. The universities had long insisted that school pupils would receive greater benefit from a more rigorous study of physical science as a preparation for university entrance; this unfortunately places an emphasis on the analytical part of engineering and ignores the important role of synthesis in engineering education. Engineering or technical drawing is fairly widespread, but its study is largely confined to the less academically able students.

The introduction of the 16–18 Action Plan by the SED—a system based on 60-hour modules—can be studied by school pupils and students at technical colleges and it is hoped that its introduction will lengthen the awareness of school pupils of the political benefits of studying engineering.

On a historical note, reference must be made to the role of Allan Glen's School in Glasgow in engineering education in Scotland. For over a century this school, which for a time was linked with Anderson's Institute, attracted well-qualified students to study engineering and science from all over the city of Glasgow and the surrounding district. Until the 1960s all prospective entrants to university in Scotland had to pass the school-leaving certificate with passes in five subjects which included passes in English at the higher grade and a foreign language. The majority of pupils at Allan Glen's offered passes on the higher grade in arts, physics and engineering and physics and chemistry, and have proceeded to university to read for degrees in one of the branches of engineering or science. This school was unique in the Scottish Educational System and the emphasis placed on a broad education including science or engineering produced a significant number of distinguished scientists and engineers, including Sir Monty Finneston and Lord Todd of Trumpington. Unfortunately, during a reorganization of the education system in the City of Glasgow, Allan Glen's School reverted to being a territorial high school and this unique educational component in directing well-educated pupils towards the study of engineering was terminated.

Case-study for Venezuela

Dr Edgar Ricardo Yajure,
Director of the Centre for Sociotechnological Transformation (FORMA), Caracas, Venezuela

Contents

A critical review of the development, the problems and the future of technological education in Venezuela

Introduction

This study analyses the historical development of the engineering and engineering-technician education system in Venezuela. The principal responses offered by this system to the changing stimuli of its environment are examined. Likewise, the possibilities of development of the same system are explored in the light of the most important contradictions revealed by the analysis. The historical epoch considered consists of approximately one hundred years, with emphasis on the study of the evolution of the system during the present century.

This epoch is examined in three stages. The first considers the antecedents of technological education during the periods of Spanish colonization and the struggle for national emancipation, as well as the process of formal structuring of the system during the last two-thirds of the past century. The second, which begins with the petroleum exploitation in Venezuela during the first two decades of this century, studies the consolidation of the system and its principal interactions with determining factors in the environment. The third stage begins with the advent of the democratic regime in 1958, in which the engineering and engineering-technician education system experienced accelerated growth and institutional diversification due to the education policy and financial resources of the state.

During the course of its development, this system has accumulated several important internal contradictions, which have begun to emerge at the present moment, when the economic well-being of the country is not improving. The most significant of these contradictions derives from the divorce between the technological-education institutions on one hand, and the enterprises of the productive sector on the other. Both have grown with the help of the state, but without establishing mutual ties. The result has been the isolation of the academic sector and a dependence on imported technology by industry.

This analysis is intended to determine the principal factors and historical roots that have caused this divorce. At the same time, it tries to show that there are elements in the present situation that could be developed so as to produce a better relationship between the two.

HISTORICAL DEVELOPMENT OF TECHNOLOGICAL EDUCATION

For the purposes of this study, the evolution of Venezuelan society can be split into three stages. The first lasted approximately three centuries, when Venezuela was a thinly populated rural society. Technology made a very limited impact and any efforts to train professional personnel were minimal. Only at the end of the past century was a modest civil-engineering education system introduced. Any significant antecedents can be found only during the second half of the eighteenth century.

The second stage lasted some fifty years and began around 1908. The petroleum exploitation by the multinational consortia is one of the most relevant economic and technological facts of this period. The petroleum activity generated, on one hand, demand for professional personnel, and on the other, fiscal benefits that permitted the broadening and diversification of the education system, and particularly of technological education. During this interval technical and engineering studies in the capital were established and extended to the two most important cities in the Western part of the country.

The emergence of the present democratic regime marked the beginning of the third stage. Here, the advantages of the high fiscal income derived from petroleum, and of an associated official policy, increased the access of the growing urban population to diverse levels and fields of education. Technological education received preference, its structure was diversified, its operations grew rapidly and were extended to the more populous regions of the country.

Each of these stages consists of various periods that are clearly distinguished to facilitate the analysis. Generally, each period is introduced with an examination of the behaviour of the most relevant factors in the environment of the technological education system; this is followed by the description and discussion of the principal changes occurring in the same system, with emphasis on the qualitative aspects; finally, a schematic representation of the state of development of the system and an examination of its quantitative evolution is given.

The first stage of development (up to 1908)

In this early stage, which ends around 1908, two periods will be examined. The first corresponds to the years before 1830, during almost three centuries of Spanish colonization, including three decades of struggle for national emancipation. During this period some antecedents of technological education will be explored. The second corresponds to the time of the caudillos, during which engineering education was introduced into Venezuela.

THE ANTECEDENTS OF ENGINEERING EDUCATION

Venezuelan colonial society was characterized by a rigid stratification of its social sectors, a strong racial discrimination and a weak economy compared with other Spanish colonies. A small group of whites born in Spain, who represented locally the absolute power controlled by the Spanish monarchy, occupied the summits of this society and enjoyed numerous privileges. The principal productive activity of this period was the exploitation of tobacco, sugar-cane, coffee and, above all, cocoa plantations, through the use of slave labour imported from Africa. Mineral extraction of gold and silver, the resources most coveted by the Spaniards, had little significance and for this reason the region never achieved the status reached by other Hispano-American colonies, such as Mexico and Peru.

In this society, engineers were almost exclusively military officers born in Spain, with the social status of nobles and the post of 'Officers and Servants of the King'; they had almost total power of decision in the execution and even financing of their duties. Design and construction of roads, churches and, principally, military fortifications were some of their most outstanding activities.

In such a social context it is understandable that there was no interest on the part of the crown in establishing technician or engineering studies. From the sixteenth to the eighteenth centuries, only the formal creation of a Geometry and Fortification Academy in 1760 is known. During a few, although undetermined number of years, this academy contributed to the technical preparation of officers and cadets of the Engineering Battalion.

Around 1800, repeated efforts failed to establish a Chair of Mathematics in Caracas University and two mathematics schools were created: one in the capital and the other in the eastern part of the country. In these schools, arithmetic, algebra, geometry, trigonometry, topography, drafting and construction techniques were taught. Personages such as Antonio Jose de Sucre, who years later became one of the principal fathers of Independence, attended these schools.

Among the factors that influenced the development of the movement for national emancipation, it is worth making special mention of the loss of the world economic leadership by Spain to powers such as England, France and North America; and, above all, the discontent created because of the obstacles impeding access to the principal political and military posts of the colony, imposed on the 'white creole' population.

The 'white creoles', were descendants of Spaniards but born on American soil, who had acquired wealth and social status from the increase in production and export of cocoa during the last decades of the eighteenth century. With this pre-eminence, the situation of inaccessibility to the key posts became very irritating. Among such posts were those of engineers, and for this reason the creole sector became fully conscious of the value of the engineers' title, and therefore, of the mathematical studies and technical education of engineers.

Towards the end of 1810, a little after the first decisive events in the struggle for Independence, the Military and Mathematics Academy was founded in Caracas. Later, in 1827, under the rule of Simon Bolivar ('The Liberator') the first Chair of Mathematics in Caracas University was created.

The ideas of Simon Rodriguez, mentor of Simon Bolivar, about the necessity to link education with work and to form qualified personnel for mechanical and agricultural tasks, only enjoyed the acceptance of the most advanced and far-sighted thinkers, who also dreamt about the creation of a big South American Republic. Such ideas were never implemented in the aftermath of the bloody episodes of the War of Independence.

THE IMPLEMENTATION OF FORMAL ENGINEERING STUDIES DURING THE CAUDILLIST PERIOD (1830–1908)

From 1830, at the end of the War of Independence, a period of violent disputes began for the control of the political power recently gained. Protagonists in these disputes were the big land proprietors, who often became regional caudillos (leaders of armed rural masses) or, at least, supported them. The predominant economic regime was characteristically almost feudal, with the latifundium[1] the primordial form of property.

By nature, the activity of the latifundia did not demand the contribution of engineers. Nor did the incipient artisanal manufacturing industry dedicated to the manufacturing of food, drinks, shoes, etc., demand the services of engineers. On the other hand, the construction of sumptuous civil works in the principal cities, generally inspired by the style of the great public constructions of the European capitals, and conceived to please the caudillo in power, did require these services. To a lesser degree, the engineers also participated in the design and construction of roads, bridges, railroads, and irrigation and canal works.

In 1831, the Mathematics Department of Caracas University was elevated to the category of academy and granted relative autonomy in relation to the university, so that, at the same time, it could serve as a military school. In this academy, the first formal structuring of engineering studies in Venezuela was effected, to satisfy the demand for

1. Property regime based on the exploitation of enormous extents of land, dedicated to agriculture and/or livestock breeding with rudimentary techniques. There are still traces of this regime in the country.

professionals. Such studies lasted six years, grouped in three two-year courses, at the end of which the graduates were promoted to the rank of Engineering Lieutenants. In 1837, the first group of these professionals, often considered the first Venezuelan engineers[1] educated in the country, graduated. In earlier years there had been various outputs of Land Surveyors, as this qualification could be acquired on finishing the first two-year course.

The boundary between military and civil engineering studies was effected in a progressive manner, keeping pace with the evolution of the differing specializations and responsibilities assumed by these professionals. Most engineering work dealt with the design and construction of simple urban buildings, as well as roads and bridges from the capital to the neighbouring towns. However, during the last quarter of the century, civil works of greater scope were built, including various palaces for the government, buildings for public spectacles, and a number of roads and railways.

In 1854, a Decree was promulgated establishing the formal separation of courses leading to the titles of civil and military engineer from those of the Mathematics Academy. In 1860, through a new decree, prerequisities for entry to the academy were established and, for the first time, the obligation to pass in subjects such as differential and integral calculus, mechanics, strength of materials and techniques in civil construction was established as a necessary qualification for the civil engineering degree.

In 1872, civil engineering and land surveying studies, still with the regime of three two-year courses, were incorporated into the Central University of Venezuela (previously called Caracas University) where two years later the Faculty of Exact Sciences was constituted. In 1883, during a university reorganization civil engineering studies began to be taught in the Faculty of Philosophical Sciences, with a duration of four years. As requisites for registration, high school and land surveyor diplomas were requested.

With its incorporation into the university, engineering studies gained in social status, but were also impregnated with all the university academicism inherited from the colonial age, expressed in such ways as the obligatory studies of antique languages and the presentation of prolonged oral exams according to rhetoric schemes.

The above-mentioned situation generated pressures which led to the creation, in 1895, of an Engineering School independent from the university. There, a four-year curriculum was introduced leading to degrees in architecture and civil and military engineering.

The curriculum of this school was the first with a modern outlook in Venezuela. In the case of civil engineering, for example, a relatively

1. Nevertheless, this nomenclature is debatable since very little is known of the contents of the curriculum used and there is evidence that there were no age or previous education requirements for entry to this academy.

broad scientific mathematical education, as well as a technically specialized education, was contemplated. The limited practical activity in projects, workshops and laboratories, however, determined that the teaching imparted continued to be by rote and of a bookish nature.

Even by 1908 engineering education was essentially limited to a single programme, civil engineering, with only one institution in Caracas. It was only complemented with the education of surveyors in Los Andes University in the western part of the country. The magnitude of enrolment of engineering students and graduates was still very small, as was the enrolment in the elementary and secondary schools.

If we add that, at that moment, Venezuela was an exclusively agricultural country, with 85 per cent of its 2.4 million inhabitants in the rural areas, it is easy to understand how little was the importance given to engineering education, and why such that there was was aimed at the field of urban civil construction. It is estimated that at that time the number of engineers in the country was approximately 200, or 1 per 12,000 inhabitants.

The second stage of development (1908–58)

This second stage comprises a fifty-year period, between 1908 and 1958, when Venezuela experienced a number of substantial changes under the influence of the petroleum activities. It was a period with petroleum but without democracy. Three fundamental periods will be distinguished.

ÉLITISM AND ENGINEERING EDUCATION DURING THE GOMEZ DICTATORSHIP (1908–35)

Two especially relevant processes, in relation to this study, took place in Venezuela in the course of the twenty-seven years of dictatorship of Juan Vicente Gomez, established in 1908.

On one hand, an economic–industrial complex was set up in the country by foreign consortia in order to exploit petroleum. Under its influence the financial and economic positions of Venezuelan society began to improve rapidly. On the other hand, in order to confront the powerful regional caudillos the dictator imposed a high degree of centralization and political administrative control of all the activities of the country, particularly of education.

Exploration, extraction and transport activities linked to the petroleum business, although conducted by foreign professionals, nevertheless generated an important demand for peripheral engineering services, especially in the design and construction of roads. The petroleum companies introduced the use of tractors, as well as diverse motorized machinery for the construction of modern roads. Soon these companies became the principal employers of Venezuelan civil engineering and surveying professionals. In the branches of mechanical, electrical and geological engineering, the absence of native engineers became evident, and added to the already chronic scarcity of agricultural engineers and agronomical specialists.

The national government also gave its own impetus to the construction of roads between the capital of the republic and some of the principal regions. This formed part of its centralization efforts and benefited from the bigger fiscal income derived from the petroleum activity. For the first time, a start was made on the creation of a national road network.

The national engineering education system was not capable of responding to the demands of its new environment. Its development was firmly restricted by the political administrative impositions of a regime that conceived education as a social privilege, and the university as a source of local power that the central government had to subdue. In 1912, engineering studies, together with other higher education disciplines, were suspended under the pretext of reorganization. Four years later, studies were resumed with the creation of the School of Physics, Mathematics and Natural Sciences, independent of the Central University of Venezuela (UCV), which was still closed. In this new school, the programmes of land surveying, architecture and civil engineering were again implemented. The award of degrees, including the administration and control of examinations, was placed under the jurisdiction of the Ministries of Internal Affairs and Public Instruction. In 1922, however, on the reopening of the UCV, the school was incorporated into the University, under the new name of the School of Mathematics and Physical Sciences.

With the intervention of the government in the area of engineering education, the latter's élitist character was reinforced. Through regulations emanating from the Ministry of Public Instruction, the criterion was instituted that the first year of engineering studies should be used to weed out those students who supposedly lacked scientific vocation. Hence during these first years only 25 to 30 per cent of admitted students progressed to further studies. In 1924, the engineering degree was replaced by that of Doctor in Mathematics and Physical Sciences.

During this period, the School of Mathematics and Physical Sciences did not diversify its curriculum or integrate into the processes of economic change arising from petroleum exploitation. Neither the students' discontent nor the complaints of high personalities linked with the Association of Engineers were sufficient to provoke substantial progress in engineering studies.

Nevertheless, some interesting achievements should be noted. Greater emphasis was given to the application of knowledge in some subjects such as topography, strength of materials, or engineering materials. The subject 'Projects of Engineering Works' was incorporated during the third and fourth years of the curriculum. Also, a second School of Physical and Mathematical Sciences was opened in 1932 in the University of Los Andes (ULA).

The scheme of engineering education which prevailed in Venezuela around 1935 did not differ very much from the one existing in 1908, except for the name of the granted degree, and a small increase in the numbers of the students enrolled and volume of graduates.

Although by 1935 there was a decrease of the traditional agricultural activity in Venezuela, and the economy had begun to follow the pattern set by the expenditure of the central government, the majority of the population of 3.3 million inhabitants lived in the countryside. Of a total of 384 professionals inscribed in the Association of Engineers in 1935, or 1 for 8,600 inhabitants, more than 96 per cent were civil engineers or surveyors.

LINKING OF TECHNOLOGICAL STUDIES WITH NATIONAL NEEDS (1936–48)

After the death of Gomez at the end of 1935, the economic, social, political and cultural regimes erected under his rule began to tumble, and, simultaneously, an intense process of change arose, marked by pressures for more democracy.

The earlier cosy relationship with the foreign petroleum consortia gave way to a policy favouring a higher financial participation in the petroleum business. This was expressed by taking certain measures, such as reserving rich hydrocarbon areas for direct exploitation by the Venezuelan State, the ending of new concessions, and the demand for national refineries. The state took great interest in the technological aspects linked to all these activities, which stimulated the development of the national engineering capacity and, as will be seen, of the technical and engineering education system. Agricultural development received a vigorous impetus. Significant irrigation works and projects, aimed at the better utilization of the national agriculture and cattle-breeding resources, were realized.

Due to the interruption in the supply of imported goods, caused by the Second World War, a process of autonomous industrialization was stimulated, based on the utilization of local raw materials and abundant, although almost untrained, native labour. Close ties between the agricultural and industrial sectors created a local agro-industrial sector. The development of a private metallurgical industry was pushed forward.

Through the combined action of physicians, investigators and sanitary engineers (the latter having graduated overseas) it was possible to overcome malaria, a disease which had caused countless deaths in the population during the preceding decades. The construction of concrete canals and the use of insecticides proved to be the best weapons. The eradication of malaria cleared the way for the occupation of the plains in the south of the country, until then very sparsely populated.

Ambitious construction projects for the social needs of the middle and low income sectors were started. Aqueducts and drinking water storage systems were constructed in the principal cities, as well as systems for collecting and disposing of sewage wastes. The construction of a modern university campus in the capital of the republic was undertaken, as was the construction and paving of roads based on modern techniques and standards. A vast national electrification programme began. The petroleum multinationals constructed the two biggest refineries in the country; and in 1947 an enormous reserve of iron ore was discovered in the Guayana region of the south.

The response made by the education system to the demands of this modernization programme and also to basic social needs, was remarkable. In the period of only thirteen years the percentage of fiscal income destined for education was doubled, reaching 12 per cent. Registrations in primary schools, as well as in higher education, rose four times. The technical and engineering education system was transformed, with the aim of adapting it to meet modern requirements. In 1937 an Institute of Geology was created in which geological engineers began to be educated with a four-year curriculum. This institute was incorporated in the Central University in 1940 and became part of the Engineering School in 1944.

Also, in 1937, higher agronomical education was initiated at the Superior School of Agriculture and Zootechnics. Its first students graduated in 1943, and in 1948 it became part of the Central University (UCV).

In 1944, a substantial reform in engineering studies took place. The annual regime was changed to a two-semester system. A programme of credits, priorities and elective subjects was introduced, giving flexibility to the former rigid programmes. Full-time appointment of academics and improvements in teaching methods were stimulated. The programme of Hydraulic and Sanitary Engineering was created. Mining and petroleum engineering studies, which were added to geological engineering, were introduced. Industrial engineering was established with two options: mechanical and chemical. The university granted its own engineering degrees again in the respective specialities. The first-year pass rate rose from 50 to 80 per cent of admitted students.

During the three-year period 1945–48, when, after an agitated political process, the first truly democratic elections in the history of Venezuela were effected, technical and engineering studies were reformed with the aim of adapting them to national requirements. In 1946, the University of Zulia, in the north-western corner of the country, was reopened and thus the number of universities offering engineering studies increased to three. These studies took place in the Engineering Schools. These, together with Schools of Architecture and Sciences, were re-organized into the Faculties of Physics, Mathematics and Natural Sciences.

A common first year for all fields of engineering was established. The importance of laboratories and workshops was emphasized. The studies of mechanical, chemical and electrical engineering were reformed in 1947, with a common curriculum for the first two years. In 1948, the programme was extended by one semester. Studies of forestry engineering began in the University of Los Andes, and petroleum engineering was incorporated in the University of Zulia. Industrial technical institutes were created to prepare middle-level technicians. The Ministry of Education emphasized the popular and democratic character of education, and recognized the necessity of integrating educational planning with the plans of the Ministries of Health, Agriculture, Development, and so on.

The state of development of the technological education system in Venezuela in 1948 was very different from that in 1935. The following two main differences might be noted. First, the range of professional activities covered was diversified, as much in the specialized fields as in relation to the types of activities; thus, the requirements in design, operation and maintenance of industrial and agricultural systems began being satisfied. Second, the enrolment of engineering students, as well as the size of departmental staff, rose nearly four times and there were also corresponding and outstanding increases in the enrolment of elementary, secondary, and middle-level technician education.

Table 1 shows the different options in engineering studies in Venezuela at the culmination of the reform process of 1936–48. Table 2 gives the

TABLE 1. Options in engineering studies in Venezuela in 1948 by branch of specialization and region

Specialization	Region			
	Capital	Andean	Zulian	Total
Civil engineering, hydraulic and sanitary engineering, land surveying	1[1]	1[1]	1	3
Agronomical engineering, forestal engineering	1	1	1	3
Geological engineering, geology, mining engineering, petroleum engineering	1	—	1	2
Industrial engineering[2]	—	—	—	—
Mechanical engineering	1	—	—	1
Chemical engineering	1	—	—	1
Metallurgical engineering[3]	1	—	—	1
Electrical engineering	1	—	—	1
TOTAL	7	2	3	12

1. The only existing options up to 1936.
2. This course existed during the 1944–46 period with two options: mechanical and chemical. In 1947 the independent studies of mechanical and chemical engineering were created.
3. Subjects of this specialization were in the Department of Geology, Mines and Petroleum of the UCV.

TABLE 2. Evolution of the engineering education system, 1935–48

Year	Schools or departments	Enrolment			Staff[1]	Graduates		
		Civil	Agronomy	Geology		Civil	Agronomy	Geology
1935	2	234	—	—	20	—	—	—
1936	2	224	—	—	20	63	—	—
1937	4	238	…[2]	…	19	—	—	—
1938	4	273	…	…	20	19	—	—
1939	4	320	…	23	25	25	—	—
1940	4	298	…	…	27	28	—	—
1941	4	301	…	50	25	45	—	—
1942	4	350	…	53	28	54	14	13
1943	4	336[3]	…	36	36	—	…	…
1944	5	403	…	…	54	22	…	…
1945	5	350[3]	…	…	58	91	…	…
1946	6	…	…	…	…	…	…	…
1947	7	743[4]	117	…	…	…	…	…
1948	10	…	…	…	…	…	…	…

1. Faculties of Geology and Agronomy are not included.
2. Data not available.
3. This number includes the enrolment in industrial engineering.
4. This number includes the enrolments in mechanical, chemical and electrical engineering.

corresponding data about student enrolment, as well as the number of graduates, and number of academic staff.

Towards 1948, the Venezuelan population reached some 4.75 million inhabitants, of which half lived in the countryside. Estimates for 1948 indicate that there were some 1,200 engineering professionals (or about 1 engineer per 4,000 inhabitants), 80 per cent of whom are were civil engineers.

DICTATORSHIP, ENTREPRENEURS, AND FLUCTUATION IN TECHNOLOGICAL EDUCATION (1949–57)

Only a few months after the democratic elections, a new military coup d'état put an end to many of the aspirations of the Venezuelan people. This regime revived and even intensified the practices of political repression of the Gomez dicatorship, and new interventions and closures of the universities and engineering schools took place.

Nevertheless, during this political regression the state enjoyed a financial prosperity without precedent and ceased to be a simple supporter of private economic initiatives. It became an entrepreneur with its own interests. This new state style stimulated many national technological activities, above all in the civil engineering construction area.

This prosperity had its origin—like most of the fluctuations in the Venezuelan economy in this century—in the circumstances of the world petroleum market. The high fuel demand generated by the reconstruction of Europe during the post-war era, and the Korean conflict, together with the Middle East crisis and the closing of the Suez Canal, converted Venezuela into a privileged source of petroleum for the West.

The sharp increase in the production and exports of crude oil took place after there had been, during the 1943–48 period, an increase in taxes levied on the petroleum companies. Thus the military dictatorship became the beneficiary of these changes, for between 1947 and 1957 the price of petroleum increased by more than 90 per cent and the production of petroleum more than doubled. Further fiscal benefits came from the licensing of new oil concessions and during the same period the revenue from oil increased four times, accounting in 1957 for about 58 per cent of total tax revenue.

This financial well-being produced a new conception of the role of the state. Now, the state, directed by the military, would lead national industrial development and create the necessary infrastructure. Thus the first state-owned petrochemical complex for the production of fertilizers and chemicals was built in the central region of the country. Also a state-owned iron and steel industry was built up in the Guayana region.

Considerable emphasis was given to the construction of major highways and roads. A large reservoir was built in order to irrigate cereal farms. Plans for further supplies of electricity continued to be implemented and a hydro-electric station was built. Many government buildings in the capital were constructed, as were several large housing projects in the capital and other urban centres. The formation of private civil engineering and construction firms was facilitated by the government.

However, the economic policies of the military regime had limitations and disadvantages. The impulse to develop agriculture and agro-industry died away. The railway system was dismantled in 1945 and replaced by roads for automobile transport. There was much construction of prestigious buildings and proportionally less attention paid to the provision of buildings for development of the country. The dependence on imports, even for basic products, increased. Contracts between the state and private firms depended largely on influence, favouritism and improvisation.

Quite generally, economic and industrial development became more and more dependent on the income from petroleum and less on the self-sustained activities of the people. There was a belief that a rich state with a power to import could be a viable substitute for a sound industrial base and well-trained workers, and indeed would provide social well-being for the population. This illusion, together with provision of urban housing, led to an uncontrolled migration from the countryside to the capital, and subsequently to the contemporary evil of massive unemployment in the capital city, and other cities, and social deprivation for large numbers of people.

Unfortunately the idea that the income from petroleum is sufficient to solve economic and social problems and sustain development is now a sort of addiction in the country from which recovery is slow.

During the military regime education was relatively neglected, with its proportion of government expenditure falling from 12 to 4.8 per cent. The need for people with technological qualifications was mostly satisfied by immigrants, who numbered 400,000 and mainly came from Europe. Thus engineering education experienced a setback in a qualitative sense, but enrolments did increase, as did academic staff and graduate output. The growth was mainly in civil engineering.

In 1951, after student unrest, the government intervened and closed the Central University of Venezuela (UCV) for two years. Students were dispersed and many went abroad to study. The most affected groups were those studying mechanical, electrical and chemical engineering, as well as mining and petroleum engineering.

The lack of agricultural development caused unemployment of agricultural engineers after 1955, in spite of their small numbers, and new enrolments dropped.

In 1953, when UCV was re-opened, the old Faculty of Mathematics, Physics and Sciences was divided into faculties of engineering, architecture and urban studies. New regulations also stimulated private higher education and two new private universities were founded with engineering as a study programme. Higher education was no longer free, and the course structure was changed. For example, the credit system was suppressed and the courses extended to five years. Also petroleum engineering was transferred from UCV to the University of Zulia. Hydraulic and sanitary engineering was eliminated at UCV and three options established in the final year of the civil engineering course. These were hydraulic and sanitary engineering, road engineering, and civil constructions. Three-year programmes were established for land surveying, building, meteorology and hydrology.

In 1956 students of the natural sciences were separated from the engineers and organized into a new Faculty of Sciences at UCV. UCV also started courses in metallurgical engineering in 1957.

Middle-level technical education actually made remarkable progress during this era. A second cycle two-year course was created which led to a diploma for mechanical, electrical or chemical technicians.

Thus at the end of the period of military dictatorship there were five-year (long cycle) courses for engineers in universities, and three-year courses (short cycle) in the public engineering schools. There was a second cycle of two years for some technicians and substantial increases of students in all groups.

Tables 3 and 4 present some quantitative data, for the period 1942–57, concerning student enrolment and graduate output. In 1957 the population was 6.62 million, of whom now only 53 per cent were in rural areas.

TABLE 3. Engineering enrolments, 1948–57

Year	Schools or departments	Enrolment						
		Civil	Agronomy and forestry	Geology, mines and petroleum	Mechanical	Chemical	Electrical	Basic cycles
1948	10	...[1]	...	...	...	...	...	...
1949[2]	10	287	...	...	...	...	...	...
1950[2]	10	360	...	...	...	...	...	...
1951[2]	10	400	45	...	...	...	...	...
1952[2]	10	579	102	...	...	...	...	...
1953[3]	12	593	255	19	11	4	32	190
1954[3]	12	653	214	50	13	4	30	580
1955[3]	12	726	194	65	12	4	35	530
1956[3]	12	754	134	111	15	10	57	843
1957[3]	12	882	186	136[4]	9	16	56	796

1. Data not available.
2. Data corresponding to UCV are not included.
3. The enrolments by branch of specialization are estimated.
4. Including five students of metallurgical engineering.

TABLE 4. Engineering graduates, 1942–57

Year	Graduates						
	Civil	Agronomy and forestry	Geology, mines and petroleum	Mechanical	Chemical	Electrical	TOTAL
1942	...[1]	...	...	...	...	...	...
1947[2]	12	...	...	...	...	...	12
1950	57	12	9	1	3	2	84
1951	51	5	2	—	—	2	60
1952	28	8	2	1	—	—	39
1953	83	32	9	1	—	5	130
1954	91	18	5	—	2	6	122
1955	116	19	5	2	1	—	143
1956	86	105[3]	1	—	—	7	199
1957	154	10	24	5	1	4	198

1. Data not available.
2. Data corresponding to UCV are not included.
3. Data not reliable.

There were 2,200 professional engineers (70 per cent civil engineers) or about 1 per 3,000 inhabitants. Urbanization affected, and indeed accelerated, the growth of technological education.

The third stage of development (1958 to date)

During the last stage, from 1958 to the present, the system of technical and engineering education has experienced considerable changes under the combined effects of a large income from petroleum and a new drive towards democracy. Three periods will be considered below.

GROWTH, DEPENDENCY AND TECHNOLOGICAL EDUCATION (1958–70)

The present democratic regime in Venezuela began at the start of the period with difficult economic and political problems to be solved. These derived principally from the decline in oil prices which continued during this period and various threats to overthrow this new regime which had replaced the former dictatorship. These difficulties were among several factors which would prevent the new leaders from re-establishing the pattern of activities which had prevailed during 1945–48 and with which they had attempted to satisfy basic social needs.

In fact, the different governments during this period used increases in revenue from petroleum production and a policy of import substitution as their principal weapons to combat social deprivation. Openings were offered to multinational firms to assemble and manufacture goods in Venezuela. Nearly 80 per cent of the profits of foreign oil concessionaires was taken by taxes and oil production was increased to its highest ever level in 1970. Import substitution bore fruit and local subsidiaries of foreign firms began assembling goods, mixing products, packing imported goods, etc. A real economic growth rate of 5 per cent per annum was achieved.

More precisely:

In the *petroleum* sphere of activity, the first state-owned oil company was formed; a programme of Venezuelization of the petroleum industry increased the proportion of local engineers employed from 20 per cent to 90 per cent of the total by 1970; desulphurization plants were built to adapt Venezuelan oil products for the North American markets.

In the *agricultural* sphere, irrigated areas for agriculture were quadrupled and flood control of the Orinoco River arranged; a reform of the land property law was envisaged but was not a complete success; the provision of a new infrastructure of civil engineering works did not produce desired effects due to the limitations imposed by cultural and educational factors.

In *transport*, road development continued: two big bridges improved communications between the Central Region and the west and south, and a network of roads was constructed around the capital; improved transport facilities encouraged industrial activities.

In *house building* and *provision of services*, throughout most of this period 1958–70 the rate of house construction declined, only reaching an earlier rate again in 1970; however, 1,500 schools were built and the construction of hospitals and the provision of water supplies went ahead, as did the generation of electricity (which tripled) and the distribution of power; a large hydro-electric station was started and telephone services were improved.

In the *industrial* sphere, foreign firms set up local plants in the Central Region using imported machinery and components. This emphasized the dependence on foreign technology and indeed did not create much demand for labour; the national petroleum and iron and steel enterprises were developed but efficiency and productivity were both very low; in the petroleum sector some firms with joint private and national ownership were formed; in 1967 a nationally owned aluminium plant was set up.

Education was given a very high priority by the state and the percentage of revenue assigned to education rose from 5 per cent in 1957 to 16 per cent in 1970. Enrolment in primary schools doubled, rising to 1.8 million in 1970, and even more spectacular was the tenfold increase of enrolments in secondary schools, so reaching 500,000 pupils. In higher education there were 80,000 students or an eightfold increase during this period. Free education was reintroduced at public universities in 1958, and academic and administrative autonomy was also granted.

In spite of the resources invested in education during this period, there was no success in reducing the dependency on the petroleum revenue, and thus on government spending, as the principal vehicles for economic development. After a promising initial reduction of the participation of the petroleum income in the ordinary fiscal budget (reduced to 52.5 per cent in 1962) this participation began increasing again to exceed 60 per cent in 1972.

During this period, six new engineering education institutions were formed: the University of Oriente (1958) in the North-Eastern Region of the country; the University of Carabobo (1958), which had been closed down at the beginning of the century, in the Central Region; the Centro-Occidental University (1962), in the Central-Western Region; the Simon Bolivar University (1967), in the capital; a new private university, the Metropolitan University (1965), in the capital; and a new type of institution, the Polytechnical Institute of Barquisimeto (1962), also in the Central-Western Region. The last three institutions were principally for the education of engineers. In the older universities, new courses were created and the practice of organizing into schools (or departments, as in

TABLE 5. Options in engineering studies in Venezuela in 1970 by branch of specialization and region

Specialization	Region						
	Capital	Andean	Zulian	Central	North-Eastern	Central-Western	Total
Civil engineering (options: hydraulics, sanitary, structural and roads), geodetical engineering, hydrometeorological engineering, land surveying, hydrometeorology	3	1	1	1	—	—	6
Agronomical engineering, forestry engineering	1	1	1	—	1	1	5
Geological engineering, geology, mining engineering, petroleum engineering	1	—	1	—	1	—	3
Industrial engineering	1	—	—	1	—	—	2
Mechanical engineering	2	—	1	1	1	1	6
Electrical engineering, electronic engineering	2	—	1	1	1	1	6
Chemical engineering	2	—	1	—	1	—	4
Metallurgical engineering	1	—	—	—	—	—	1
Systems engineering, computing	1	1	—	—	—	—	2
TOTAL	14	3	6	4	5	3	35

the case of the polytechnical institutes) for the teaching of programmes was established again.

In Table 5, the new pattern of options in engineering studies in 1970 is shown. There is a significant increase of such options and they are now better distributed around the geographical regions. Note the incorporation of the new branch of computing and systems as an option.

In 1958, teaching by semesters was reintroduced, but the five-year engineering curricula were retained. During the first six years of this period (1958–63) a return to the reform spirit of 1944 was hoped for by emphasizing (a) the linking of theory and practice, (b) manual and workshop activities during the first semesters, (c) the implementation of project activities, (d) the solving of real problems, (e) the study of the characteristics of local industries, and (f) the creation of relatively specialized options towards the end of the courses. In chemical engineering, for example, options in chemical and food industries existed with emphasis on the study of chemical processes, and design of industrial equipment.

This approach, nevertheless, was replaced in 1964 by another that emphasized a basic education of a scientific–mathematical type, and, fundamentally, left the professional training in the hands of the industrial sector. It appears convenient to analyse in detail this reform because this type of curriculum has basically prevailed until the present time in most universities.

There were two central arguments then employed to justify the reform of the curricula. In the first place, there was supposed to be a need to form a type of 'elastic', 'electric' engineer, capable of acting efficiently in design, as well as in maintenance, production, administration, and even in teaching and research. Emphasis was on a 'basic' education for future professionals.

Second, to reduce the 'technological gap' that separated Venezuela from industrially developed countries, it was considered necessary to select a curriculum permitting a type of 'pursuit by leaps and bounds'. The premise was that 'the state of development of the country should not be the limit of the level of engineering education', or that 'the level of a given curriculum should not be that corresponding to some average, but at the present state of knowledge in its field'.

Based on this argument the new curricula were designed, taking as a model the type emphasizing engineering sciences being applied in the United States, and retaining only some applied subjects at the end of the five-year courses. The concept of engineering sciences often designates such disciplines as mechanics, electromagnetic theory, thermodynamics, strength of materials, transport phenomena, etc., which serve as a direct support to the professional activities in different fields of engineering.

Other interesting aspects of the new curricula were: the extension of the common basic cycle to three semesters for all programmes; the achievement of greater flexibility by offering a higher number of elective subjects; the reduction of the number of subjects per semester; and better co-ordination between subjects, reducing the repetition of former curricula.

The author believes that the decision taken at the beginning of the 1960s to follow the North American example by emphasizing the engineering sciences to the detriment of professional education (in a strict sense), was unfortunate.

In North America the decision to move towards engineering sciences was adopted in the context of: (a) the armaments and space race which did not demand engineers having knowledge of known equipment and systems; (b) a very well-developed system of graduate education and research, capable of completing the basic education of undergraduates; (c) great experience, accumulated during the course of more than one hundred years, of engineering educational curricula leading to the practical employment of knowledge; and (d) an efficient system of industrial training for the new graduates.

The Venezuelan antecedents were the opposite: (a) a great number of urgent, unsatisfied, basic needs (nutrition, housing, etc.) whose attention by engineers did not demand spectacular innovations, but rather the adaptation of technologies already developed in other countries; (b) the lack of graduate education, and an almost non-existent system of research and development; (c) a scarce technological experience based on only a few years of having engineering education for industry; and (d) a situation where the new graduates, almost untrained, sometimes have to assume immediately professional responsibilities. Under such circumstances, the North American type curriculum seemed inadvisable, since it did not match any of the national needs.

To a certain extent, the result of the adoption of this engineering curriculum in the principal engineering universities has resulted in a large number of professionals without the opportunity to apply their abstract knowledge. A minor exception to this has been the small number of professionals who have had the opportunity to continue their education abroad, or have acquired professional training in a few large companies linked to the national corporations. Concerning these professionals, it is of interest to mention that at last the country has begun to develop a capacity for basic design and systems innovation. Although the price paid by the nation is high, in terms of the loss of potential technological human resources, the policy of educating engineers according to standards of technologically developed countries has facilitated the access, by local graduates, to postgraduate courses in foreign countries, and, above all, in the United States. It is possible that this is the most beneficial feature of such a policy.

Another interesting development of this period, and contrary to the tendency already outlined, was the creation in 1962, based on initiatives of Unesco, of the Barquisimeto Polytechnical Institute. This institute has been dedicated to the education of professionals oriented towards productive industrial activities. Initially, a system for educating technologists or engineering technicians was adopted; later, one aimed at educating production engineers, with a higher professional technical capacity than those graduating from the universities was introduced. The number of engineers educated according to this approach, nevertheless, has always been much smaller than the number of graduates of the universities.

Thus by about 1970, technological education had spread into a large number of regions and there was a new type of institution, the Polytechnical Institute. The total student enrolment in the technological programmes increased and exceeded 18,000 students, nine times the enrolment in 1957; the number of engineering schools as well as the flow of graduates tripled during the same period, reaching a total of 650 graduates per year as compared with 200 graduates per year in 1957. However, it should be noted that this increase in the number of graduates is far below the increase in enrolment, which suggests poor performance of the system.

This growth has continued, on one hand, in response to official efforts in favour of industrial and agricultural development; but, on the other hand, it has been, above all, the result of attention to the demand for higher education by the growing urban middle classes. They have found in access to technological education a way to benefit from public employment or from a capacity to undertake contract work for state organizations.

Towards 1970, the number of engineers and related professionals (except architects) in Venezuela was close to 8,000. The percentage of civil engineers, for the first time, decreased below 50 per cent; agronomical engineers, with some 1,000 professionals corresponding to 12.5 per cent, and mechanical and electrical engineers, with close to 6 per cent each, were the next most significant specialisms. In relation to a population of approximately 10.2 million inhabitants, with only 28 per cent in the rural sector, the fraction of engineers in relation to the total population now attained 1 professional per 1,275 inhabitants.

INSTITUTIONAL EXPLOSION AND DEMOCRATIZATION OF ACCESS TO TECHNOLOGICAL EDUCATION (1971–81)

Towards the end of the 1960s there was stagnation of the national income derived from petroleum. The constant increase in the volume of production, reaching its peak in 1970, was not sufficient to offset the effects of falling prices. The interest of the foreign petroleum companies in continued increases in production fell, due to the fact that the Middle East wells had begun to give better yields.

On the other hand, the limited market of the high income social sector, capable of consuming the relatively sophisticated manufactured products assembled by the subsidiaries of the foreign companies, became saturated. At this point, there was an economic bottleneck, commonly characterized as the exhaustion of the process of substitution of imports of consumer goods.

With this exhaustion, a new economic development arose based on the national production of intermediate goods for export, principally petrochemical and metallurgical. Commercial exchange with the neighbouring Andean countries began to be considered a priority. This model was conceived as a kind of continuation, on a regional scale, of the imports substitution policy initiated years earlier on a national scale. Limitation of financial resources available for urgent investment demands, and the under-development of native human resources, were considered as the principal obstacles to be overcome.

In December 1970, the twenty-first OPEC Conference was held in Caracas, at which the petroleum exporting countries adopted a strategy against falling prices. Among other measures, it was agreed that the same exporting countries would determine the reference price of crude oil, to collect taxes from the foreign companies. As a result of the application of the new policy, and in the context of growing demand for hydrocarbons, the average reference price of Venzuelan petroleum doubled between 1970 and 1973. Consequently, the fiscal income of the state increased by some 60 per cent during the same period.

During the last days of 1973, the war in the Middle East and the embargo of exports of petroleum to the United States adopted by the Arab countries, a world energy crisis exploded. As a result petroleum prices tripled, as did the budget of the Venezuelan Government.

With the new financial resources, the state was now able to buy up both the petroleum and iron industries. With this nationalization the industrial development of the country was accelerated and amplified, with a massive importation of equipment and technology. Real economic growth between 1973 and 1977 exceeded 7 per cent.

Among other large projects in the 1970s, a great petrochemical complex was constructed in El Tablazo, new works were built which multiplied by four times the capacity for steel production; the aluminium-producing sector experienced a remarkable growth and surpassed by forty times its installed productive capacity; and capacity for the generation of electricity was quadrupled. The plant for crude oil processing in the refining industry was modernized to be able to accept a greater volume of heavy crude oils and increase gasoline production. An important machine industry was established, oriented towards the production of capital goods. Large investments were made in order to develop the agricultural sector. The labour force was almost completely employed at this time.

Alongside this industrial growth, the 1970s also witnessed an increased dependency on petrolum income. Between 1970 and 1981 the revenue from the petroleum sector passed from 60 to 77 per cent of the ordinary fiscal income and from 20 to 27 per cent of the Gross Territorial Product. During the same period, imports multiplied seven times, and the nation returned to acquiring from abroad many essential goods such as food and textiles that had not been imported in the previous decade.

The educational system benefited from the financial bonanza, since it maintained its 16 per cent share of a much higher budget. Enrolment in primary education increased more than 60 per cent, exceeding 3 million students in 1981. Secondary education enrolment doubled, exceeding 1 million students. The enrolment in higher level education was quadrupled, exceeding 330,000 students.

The need for higher industrial productivity within an economy that was hoping to compete in external industrial processes, made it desirable to renew efforts to educate more people in the technologies. Through the National Pre-enrolment System of high-school graduates for higher education, the state tried to direct the demand for higher education towards this area. Obtaining a degree in the technological field, on the other hand, continued to be a privileged route up the social scale. These two factors acted jointly to justify the expansion of the technological education system, which acquired an explosive character, due to the availability of financial resources.

Between 1970 and 1981, six public and three private universities were created, the majority almost exclusively oriented towards engineering education. Two of these new public universities were established in the less populated region in the southern part of the country, and another was inspired by the Open University in the United Kingdom. Three new Polytechnical University Institutes were created, one of them in Guayana, using as a model the Polytechnical Institute of the earlier period. Following the French model of short-cycle technological curricula, thirteen public technological university institutes and seven private ones were created, for the education of engineering technicians. Also, in three university colleges, similar to the Community Colleges of the United States, short-cycle technological programmes began to be offered. The number of options for technological degrees was significantly increased, above all in the regional areas. Numerous extensions of principal institutions were established in small and medium-size cities.

An ambitious scholarship plan, principally directed towards higher technological studies abroad, was started.

As regards curricula, the principal universities of the country generally maintained those subjects which emphasized basic education and had few connections with the industrial sector. A remarkable exception to this trend, along with the polytechnical institutes, was the efforts in the University of Carabobo, and recently in the Technological University of the Central Region (private), to link engineering education to local industrial needs and direct it to the preparation of designs and projects work.

Another effort that is worth pointing out is the impetus given by the Industry–Education Foundation to programmes of co-operative education for the industrial training of students on a national scale. Generally speaking, the recently formed universities and institutions have been more inclined to establish ties with the productive areas and welcome initiatives such as industrial student training. On the contrary, the institutions established before 1970, and above all the three older traditional universities created before 1958, have tended to remain relatively isolated from industry.

In one of the principal official documents of policy for higher education approved at the end of the decade, the fragmentation of the higher education system was proposed as a solution to the problem of quality. Part of the system was to be dedicated to the democratization of higher education, and the other to the preservation of the excellence of the educational process.

The main ways to achieve the democratization were the creation of higher level short-cycle programmes, most of them in technology, and the establishing of a regime of open studies, with the creation of the National Open University (UNA). In the Fifth Plan of the Nation, officially approved in 1976, it was revealed that, for 1980, 32 per cent of an estimated enrolment of 305,000 students at high level should enter the institutes with short-cycle programmes, and 10 per cent enter the Open University.

Although the planned enrolment in short-cycle programmes or Open University could not be achieved, total growth of the higher level enrolment surpassed the expected level. With this growth, it was possible to significantly democratize access to higher education. The fraction of the population aged between 20 and 24 years in higher education rose from 9.49 per cent in 1970 to 22.68 per cent in 1981.

At the same time, enrolment in technological education as a percentage of total higher level enrolment increased from 22.0 per cent in 1970 to 27.3 per cent in 1981.

A fundamental step in the process of development of the Venezuelan technological education system was the introduction of graduate studies. This had been conceived to prepare professionals for management, research, and development activities, as well as for improving education itself.

In spite of the great number of institutions created, the three older universities continued to absorb almost half the total number of enrolments in technical education. The Polytechnic Institutes continued to absorb only a small fraction of the total number of long-cycle programme students.

An examination of the figures on enrolment shows that short-cycle programmes have absorbed only a minor fraction of the total enrolment in higher level technological education. This is far below the objective of the Fifth Plan of the Nation, that is 32 per cent.

The number of engineers in Venezuela in 1981 may be estimated as some 30,000 professionals, of whom about 28 per cent are likely to have been civil engineers. This means that, in respect of a population of 14.6 million, with less than 20 per cent in the countryside, the proportion of engineers increased to 1 per 490 inhabitants.

THE EMERGENCE OF CONTRADICTIONS IN THE TECHNOLOGICAL EDUCATION SYSTEM (1981 TO DATE)

With the passing of the 1980s the abundance of the past decade has begun to disappear. The decline in petroleum prices has been accompanied by a decrease of demand in the world market that has not permitted an increase in production, and from 1981 on the economic parameters began to vary drastically. During three consecutive years (1981, 1982 and 1983), there was a real economic decrease.

In February 1983, the country underwent a strong currency devaluation, which caused a reduction to one-third of its purchasing power on international markets. This forced an abrupt fall of imports, which had grown constantly during the last decades. Inflation accompanied the economic recession, in a variant of the process that is known today as 'stagflation'.

Many industrial companies have gone bankrupt. Unemployment of the work-force has been calculated at some 15 per cent. This time, engineering professionals are being severely affected, having an estimated rate of unemployment of 10 per cent. The public debt contracted with the developed countries during the earlier period constitutes a heavy load that consumes almost a quarter of the state budget. Up to now, there have been no signs of effective recovery of the economy.

With this new situation, the internal contradictions of a technological education system that had evolved without establishing links with the productive sector, have emerged. The accumulated qualitative problems, which during past years remained neglected as secondary matters compared with fast quantitative growth, are now the order of the day.

In wide circles, it is taken for granted that this system required a far-reaching reform in order to match present circumstances, and to effectively combat the national paralysis. It is considered necessary to diminish the vulnerability of the economy, taking into account the inconstancy of the international petroleum situation. However, it has not been possible to define the role of the technological education system to achieve these goals. The inertia and practices inherited from the past are serious obstacles to be overcome.

Venezuela received during the last decades, and above all in the 1970s, a high volume of income which was not generated by the local productive work-force and which did not imply political sacrifices for the population. A considerable sector got accustomed to the idea that the solution to their problems did not depend on their own work and training, but that it was always provided from above, from a rich state. They expect that at any moment petroleum prices will increase and that everything will again be 'normal'.

However, it is unlikely that such 'normality' will ever return, and sooner or later the country will have to face reality. The technological education system, if it wants to overcome this situation, will have to solve its own problems.

Dilemmas inherited from the past, and the future of technological education in present-day Venezuela

Historically, Venezuelan technological education has achieved remarkable progress and has acquired considerable dimensions. However, it has also accumulated contradictions that up to now have not been solved. These can be associated with the interaction that the system has established with its environment, the type of curricula that prevails in the principal engineering education centres, and the efficiency or internal performance of the system.

Until a few years ago, the system enjoyed good financial backing from the state that permitted it to grow, although without paying attention to its contradictions. The present situation, and the best forecasting of circumstances in the near future, suggest that the time has arrived to redefine its mission to Venezuelan society. It has been sheltered for a long time in a type of greenhouse or artificial chamber, and now it must begin to function satisfactorily in a somewhat less accommodating environment.

THE CENTRAL DILEMMA: DEPENDENCY VERSUS AUTONOMY

Since the days of the caudillos of the past century, when the Venezuelan technological education system was formally structured, only weak ties have been established with the community and the productive sectors. The industrial development of the country has been supported principally by external technological expertise. Technological progress in agriculture has been slow, in spite of the under-nutrition which affects large sectors of the population. There has not been a correlation between the increase in the number of agronomical engineers, for example, and an acceptable degree of self-sufficiency in food supplies.

Developments in engineering education have always been concerned with objectives of a different nature. These have been, among others, the need to construct prestigious works during the caudillist and Perezjimenist periods; to form a privileged social élite in the days of the Gomez dictatorship; or to satisfy the social aspirations of the urban middle sectors since 1958 up to the present. Exceptionally, certain efforts have been made to forge industry–education linkages during the period from 1936 to 1948; or in the case of some regional experiments, during the last decade.

The increase of the petroleum income has been the fuel for the engine that, since the end of the 1940s, has driven industrial growth in the country. This growth has reached considerable size, especially since 1971. Venezuela has become equipped with large and modern industrial plant that exceeds the capacity of local human resources to make it function efficiently. The technological and industrial dependency of the country is closely related to this impotence to convert into productive and profitable use the industrial investments of the past.

This dependency has been highlighted by a UNIDO expert in a recent study. According to this author, Venezuela possesses equipment, processes and technologies characteristic of a developed country, but with the limitations in the labour force of a typical developing country. If this discrepancy is not corrected, Venezuela runs the risk of a complete economic paralysis.

In fact, there was, and is, a belief that the petroleum income, which had permitted the import of technology created by foreigners, could also make up for the lack of an indigenous technological capability. Present circumstances, and the economic paralysis that the country is experiencing, reveals the fragility of such a belief, and the pertinence of the warnings. In a country with an enhanced productive potential, with scant technological experience and with a great number of unsatisfied basic social needs, engineering and technical education should be placed at the service of a process of self-sustained development. At present, it is not clear what will be the response of the education institutions to the challenge posed by the problems of obtaining technological autonomy.

On one hand, there exist tendencies, in the education sector as well as in the productive and governmental sectors, that act to achieve such autonomy. In the University of Carabobo, the Technological University of the Central Region, and others, at regional levels, as well as some technical and polytechnical institutes, important experiments are being carried out to link social and productive needs. In various congresses on engineering education held recently the importance of this linkage has been underlined.

In the productive sector similar tendencies are to be found leading, for example, to the creation of the Industry–Education Foundation, already mentioned; and there are interesting efforts in the Central Region of the country. In the Guayana Region, where the principal investments of the 1970s are located, movements to participate in a search for greater technological autonomy have been generated. The State Petroleum Company, the principal one of the country, is deliberately relying, whenever possible, on national engineering capabilities.

On the other hand, however, there are sectors that exemplify the inertia of former days and the resistance to changes. The academics of the main universities of the country represent one of the more conservative forces. In the name of academic excellence, they are reluctant to participate in the process of innovation or assistance to industry. The traditional business sector prefers to continue importing equipment and foreign expertise instead of taking the risk of relying on local technological capabilities. Government policy is to go for immediate vote-winning results rather than concentrating on devising long-term technological plans.

In the coming years, the result of the debate between these tendencies will have much influence on the destiny of technological education. But now, paradoxically, perhaps the decline of the importance of petroleum as a world's energy source, will be the factor that will serve as the principal inducement for change.

ÉLITIST EDUCATION VERSUS WORK EDUCATION

The élitist conception of engineering education had its roots in the Gomez dictatorship years (1908–35). The principal objective of the engineering schools was to select the members of a socially privileged élite who would exercise administrative authority over the rest of the workers. The engineer was not seen as a highly qualified worker capable of producing technology, but as a 'Doctor', whose mere title inspired respect in the other workers, in charge of preserving the functioning of enterprises and institutions. The favourite mechanism for the realization of this selection would be subjects of abstract type such as mathematical analysis, descriptive geometry or rational mechanics. These would serve as a kind of filter to purify the student body from undesirable elements, supposedly not fitted for this vocation.

Although things have changed very much since that time, this conception is still present in the technological education system of the country. It is difficult to calculate the loss of potential human resources that the nation has undergone because of the hundreds of thousands of young people who, in the last decades, have abandoned the classrooms during the first two years of engineering studies, without having had any opportunity to show their capacity to solve problems of a truly engineering character. The reality of a professional engineering practice in Venezuela has facilitated the survival of this concept. The proliferation of enterprises and organizations established with finance from the state and using imported expertise has generated numerous administrative jobs. To these, many young engineers have been recruited without having exercised their original profession.

At present, when the capacity for generating administrative jobs by the state has diminished the employment of professionals is considered proof of the lack of selection, and, therefore, the principal cause of the saturation of engineers on the labour market. The élitist concept is reinforced by this situation in the minds of some academics.

The author considers that this saturation is only of engineers without professional experience who want administrative posts in enterprises using imported technologies. The potential demand for engineers to do productive work and to generate and adapt local technologies, on the other hand, is likely to be unsatisfied to a large degree. In a study of the market of engineering professionals in Venezuela, something similar was concluded. According to this study, with the traditional ratios of engineers in productive activities to the total work force, Venezuela was considered within the near future to have an excess of professionals. Nevertheless, if such ratios were elevated to the corresponding values in the United States then a serious shortage of engineers was foreseen, especially in mechanical engineering.

In the coming years, the nation should decide whether to continue the traditional import of technology—in which case there will be an excess of engineers in the labour market and the élitist practices will be in force—or to progress in the sense of self-sustained development. Then it would be necessary to re-educate the engineers and technicians for productive work and local generation of technology, and to look for new approaches in preserving the quality of teaching.

ENGINEERING SCIENCES VERSUS PROFESSIONAL EDUCATION

The engineering-science type of curricula imported in the 1960s by the main engineering education institutions contributed to the substantial absence of professional training in their courses. It was assumed that industry would perform this function.

However, there is reason to believe that only exceptionally has industry been able to cope with this. Some empirical research suggests that the majority of graduates do not receive any training before assuming responsibility in their first job in industry; and that this responsibility includes, as a general rule, the evaluation of projects, the elaboration of reports with technical recommendations and decision-making at a more demanding level than that of their peers in technologically developed nations. If the élitist ideas have caused losses to the nation in terms of potential human resources, the lack of professional training has caused additional ones because of the limited capabilities of graduates, who often have not received professional training at their work, to solve industrial problems.

This disappearance of professional studies from the curricula has particularly affected the discipline of industrial engineering because the educational institutions adopted such plans at a time when there were no practising industrial engineers in the country. It is very probable that this 'de-professionalization' could account for the phenomenon of engineers striving to enter the administrative sphere. They experience difficulties making practical use of their basic knowledge when faced with technological problems.

SPECIALIZED EDUCATION VERSES GENERAL EDUCATION

The arguments used to justify an emphasis on the basic engineering sciences could also be used to justify a general education, though not necessarily in engineering science. The accelerating progress in technology of today makes it useless to concentrate on the design and operation of equipment that could well be obsolete by the time a student graduates. On the other hand it should be noted that the rhythm of engineering innovation is not uniform. In the large highly industrialized nations, changes are rapid in fields connected with defence, information, energy and space but much slower in fields such as agronomy, civil engineering, metallurgy, and so on. These latter activities are more directly connected with satisfying needs such as roads, housing, clothing, nutrition and agriculture, as found in developing nations.

There are therefore strong arguments in developing countries that attention in technological education should be given most, if not wholly, to equipment and processes of use in the more traditional branches of engineering and agriculture. In Venezuela there is still a great need to inculcate efficient methods of operation and maintenance as well as design and production.

Second, a lasting engineering education need not be based on engineering sciences but can be based on professional engineering activities through attention to methods of planning, processing information, simulation and optimization, decision-making, and so on. Project work can lead to report-writing, design etc. of relevance to local industrial activities.

FRAGMENTATION VERSUS INTEGRATION

The initiatives taken after 1958 to build schools and to increase enrolment in primary and middle-level education, contributed to an increased demand for higher education facilities. The need to satisfy this demand, in a context of minimum planning and abundant financial resources, led to the creation of a number of higher level institutions.

The solution to the problem of democratizing access to higher technological education, and at the same time preserving the assumed excellence of education, has consisted in dividing the technological system into two subsystems: one for engineering technicians and production engineering education in technical or polytechnical institutes, and in the university (community) colleges; and the other for a more theoretical engineering education in the universities, this practice has in turn created more problems.

High-school graduates perceive that the majority of mass education institutes that offer short-cycle programmes are of inferior status. They do not offer the possibility of eventual transfers to long-cycle programmes and therefore are considered a dead-end. Only those who have no other alternative enrol in these options.

The universities, on the other hand, are faced with problems of low yield, repetitions and drop-outs, with the consequent waste of financial, institutional and human resources. The desired excellence is attained by few and seems to be achieved at a high cost of failure of many others.

From another point of view, that of professional practice, engineering technicians complain that, whatever capability they show, promotions are blocked because of the lack of an engineering degree; while the engineers claim that they must often assume the duties of the technicians due to the shortage of the latter.

The integration of these subsystems of higher technological education seems to be a necessity. Some moves towards this have been made, and there even exists a Presidential Decree, dated 1979; but despite the efforts made the integration reached so far has been limited. Some universities have begun to accept an engineering technician diploma as meeting some requirements for entry into engineering education, but this does not satisfy the expectations of those who make the transfers, or of the government.

A definitive solution to the dilemma of integration can be found only when other contradictions described above are also sorted out. While the linking with the industrial community of the selection of students and curriculum design is stubbornly opposed, little will be done towards true integration. On the other hand, with a coherent curriculum allowing transfer from short- to long-cycle programmes, without sacrificing time, some integration seems feasible. This curriculum might consist of three cycles. The first would prepare engineering technicians for the activities of operation and maintenance of existing industrial systems. The second would educate engineers in the design of new systems. The third would consist of graduate studies oriented towards the education of professionals for research or management of productive systems. The central theme of the three cycles would be the realization of real projects on industrial problems of growing complexity.

In a private university, the Technological University of the Central Region (UNITEC), and in the National Open University (UNA), integrated studies have already been put into practice. There, the intermediate degree of Engineering Technician is granted on the way towards the degrees in informatics, and industrial and systems engineering. In two other universities, the University of Driente (UDD), and the Simon Rodriguez University (USR), they have also experimented with educational programmes of this type. The present government is promoting studies for integration in the Guayana Region, which would be used as a pilot project, and this has created great expectations. In the Polytechnical Institute of Guayana, initiatives are also being taken in order to implement the new model.

ACADEMIC RESEARCH VERSUS INDUSTRIAL RESEARCH

Since the creation of graduate studies in engineering colleges in Venezuela many debates have been held about the type of research that should be undertaken. It is undeniable that pure or academic research is a proper activity in a university. However, in a developing country like Venezuela it does not seem desirable to copy procedures in some, but not all, industrially developed countries where universities specialize in pure research whilst industries pursue technological research and development.

Venezuelan industry, most of it created almost overnight, is still engaged in the task of operating and maintaining its equipment and processes. It has neither the will nor the resources for R&D activities. The universities and similar education centres have the best facilities for research but there is no guarantee that they can produce results that will be of use to local industry. It would seem that Venezuela today requires a new pattern of education–industry co-operation. Valuable moves towards such patterns are being made by the Education–Industry Foundation, the National Council for Scientific and Technological Research, and the Ministry of Science and Technology.

MICROELECTRONICS VERUS MACROMECHANICS

In most industrially developed countries the first engineering disciplines of importance were those related to the use of land and provision of shelter. Amongst these are architecture, civil engineering, mining and agricultural engineering. Later came metallurgical, mechanical and chemical engineering, all connected with the processing and use of materials. Later still there was electrical engineering concerned with energy, and finally those disciplines related to the processing and control of information such as electronics, informatics and cybernetics. There has been a tendency for the majority of graduates at any time to gravitate towards the newer disciplines.

In Venezuela the engineering disciplines related to land use and living space have never stopped being the most popular, nor has their social significance been surpassed by that of the newer industrial engineering disciplines. In particular, mechanical and chemical engineering, which should be central to the business of developing the abundant metalliferous areas and petroleum reserves of the country, have not attracted students as they ought. On the other hand, the precocious development of electronic and electrical engineering, and now computation and informatics, is depending on the import of technology, equipment and components rather than on local expertise and manufacture.

It seems that Venezuela wants to omit the phase of developing its own basic manufacturing industry and go from civil engineering activities to those of a 'post-industrial' type linked to energy and information. This would be a phenomenon corresponding somewhat to what has happened at the professional engineering level where the centre of gravity seems to have passed from the primary to the tertiary sector without the secondary or industrial sector ever having been a majority activity.

This is not a matter of defending linear evolution, but only pointing out the inconveniences that this evolutionary jumping style can cause and has already caused, and suggesting the existence of yet another dilemma to be solved. While in technologically developed nations the search for control and optimization of energy and information systems began after the development of a manufacturing industry, including an important metalworking sector; in Venezuela the import of many of the most simple metallic spare parts and accessories continues to this day.

It is estimated that 60 per cent of Venezuelan imports during the 1960s consisted of metallic products. With the sudden fall of the flow of imports, industrial development is beginning to face a bottleneck which is difficult to overcome. This occurs, paradoxically, in conditions in which the local metalwork industries are suffering from large idle capacity. The need to develop a local capacity for mechanical design, capable of linking the unsatisfied demand for metalwork products with the large potential market, is now urgent.

While hydrocarbons are being exhausted day after day and agriculture remains in an unproductive state, incipient local chemical engineering has not as yet been able to make any significant contribution to the petrochemical or to the fertilizer and pesticide industries.

It seems clear that with these limitations the nation cannot completely engage, like the developed countries, in new industries such as electronics without first ensuring a viable manufacturing industry for basic local products.

The policy of supporting the local engineering capacity that the petroleum industry is now sponsoring and the assignment of production of spare parts to local companies that the state-owned steel and aluminium companies are beginning to push forward can contribute very much to the progress of local mechanical, chemical and metallurgical engineering activities.

TECHNOLOGICAL VERSUS SOCIO-HUMANISTIC EDUCATION

Finally, it is interesting to emphasize that the reorientations of professional activities and the technological education system cannot progress without adequate humanistic vision. In the context of social structures of developing countries, technology exerts a greater impact than in developed countries. The participation of engineers and technicians in the social and political sphere occurs here with more frequency and earlier than in the case of their colleagues in developed countries. This is even more so if it is considered that, in Venezuela, the major part of the bigger enterprises belong to the state, and have been created with finance derived from resources belonging to all Venezuelans.

The motivation to overcome the technological dependency of the country, as well as that required to promote the transformation of the technological education system, cannot be nourished from a better source than a clear comprehension of the social and human problems of Venezuela, Latin America and mankind in general.

Conclusion

During the course of the last five decades, and especially during the last fifteen years, the system of Venezuelan technological education has experienced a considerable growth and has diversified the educational options offered in many disciplines and also provided most regions with engineering institutions. Venezuela now has an infrastructure that permits the education of engineers and technicians on a large scale. In spite of this progress, technological education in the country suffers from evils that need to be corrected in the near future.

The first and principal of these is the failure to link engineering education with the real needs of the community and to the productive sectors. The second is probably the failure of these institutions to co-operate and integrate. Lack of attention to the most demanding problems of the community, agriculture, and industry; the dominance of academic preferences in teaching and research; and the inefficient use of the available resources for technological education are indicators, and consequences at the same time, of the existence of such evils.

The exploitation of petroleum has permitted the Venezuelan State to support an accelerated process of urbanization and create a large free engineering and higher technician education system, in which local industrial development has found neither appropriate support nor involvement.

This situation, which has changed the face of Venezuela, has been subsidized and protected by the state and based on the indiscriminate import of foreign technology.

The recent process of social democratization has increased substantially the demand for access to higher level technological education. The latter was conceived as a privileged route for social promotion and for enjoying the benefits derived from petroleum riches.

The combined effects of the two factors has resulted in an accelerated expansion of the educational enrolment and the number of technological disciplines in conditions in which the qualifications of graduates do not satisfy the expectations of industry and agriculture. The most recent consequence of this has been the phenomenon of massive unemployment of engineers. This occurs, precisely, at the moment in which the decrease of the purchasing power of the national currency is exerting pressures in favour of the national production of goods once imported.

Present predictions suggest that in the near future the declining purchasing power of the petroleum income will continue, with the consequent difficulties in importing goods and technologies for the industrialization process. More restrictions will be applied in order that the state can use more of its resources to satisfy urgent needs for housing, food, etc.

In such a situation, the technological education system seems to have three main options. The first would be to maintain the status quo, trying to survive until the world petroleum market improves and everything returns to 'normality', as in the past decades.

Another would be to try to look for a solution based on the experiences of past periods, and throw out democratization of technological education. In this way, the 'élitist' conceptions, which see the engineer not as a producer of technology but as a member of a hierarchy in a technologically sterile ambience would be once again imposed.

The third would be to accept the challenge of present circumstances and try to put the infrastructure of technological education in harmony with industrial technological and economic development. In this case, it would be necessary to undertake a comprehensive analysis in order to establish the professional profile of the engineers required and the curricula needed in the engineering courses, the object being to produce technologists capable of exploiting the resources of Venezuela for its citizens.

Bibliography

ARAUJO, Orlando. *Situación industrial de Venezuela*. Caracas, Ediciones de la Biblioteca de la UCV, 1969. (Serie Nuevos Planteamientos, No. 1.)

ARCILA FARIAS, Eduardo. *Historia de la ingeniería en Venezuela*. Caracas, Colegio de Ingenieros de Venezuela. 2 vols.

BOLÍVAR, Simón. Método que se debe seguir en la educación de mi sobrino Fernando Bolívar. *Simón Bolívar, Obras completas*. Vol. 3, pp. 837–9. Compilación y notas de Vicente Lezuna. Caracas, Ministerio de Educación Nacional.

CASTILLO PINTO, Enrique. *Estado actual de los estudios de ingeniería a nivel de pregrado en Venezuela. Trabajo presentado en el Congreso Mundial sobre Educación de Ingeniería, Chicago, 1965*. Caracas, UCV, 1965.

COLEGIO DE INGENIEROS DE VENEZUELA (CIV). *La ingeniería, la arquitectura y profesiones afines ante el proceso de desarrollo de Venezuela. Trabajo presentado en el 8.º Congreso Venezolano de Ingeniería*. Caracas, CIV, 1969.

LEAL, Ildefonso. *Historia de la Universidad Central de Venezuela*. Caracas, Ediciones del Rectorado de la UCV, 1981.

Leyes y decretos de Venezuela, No. 18 (1830–40), No. 3 (1851–69), No. 5 (1870–73). Caracas, Biblioteca de la Academia de Ciencias Políticas y Sociales. (Serie República de Venezuela.)

MUDARRA, Miguel Angel. *Historia de la legislación contemporánea en Venezuela*. Caracas, Publicaciones Mudbell, 1978.

Recursos humanos en Venezuela; estudio preliminar de profesiones prioritarias. Cambridge, Mass., MIT Centre for Policy Alternatives, 1976.

SALCEDO BASTARDO, J. L. *Historia fundamental de Venezuela*. 5th ed. Caracas, Universidad Central de Venezuela, 1976.

TOVAR, Ramón A. *Venezuela, país subdesarrollado*. Caracas, Ediciones de la Biblioteca de la UCV. (Colección Avance, No. 6.)

YACKOVLEV, Vladimir. *Engineering Education in Venezuela*. Annual Meeting of the ASEE, Southeastern Section, American Society of Engineering Education, 1967.

YAJURE, Edgar. *Encuesta sobre los antecedentes, la carrera y perfil profesional y la formación académica de los ingenieros mecanicos y profesionales afines en la Región de Guyana*. Ciudad Guayana (Venezuela), Instituto Politécnico de Guyana (IUPEG), 1983.

YAJURE, Edgar. Por un cambio sustancial de los actuales planes de estudio de ingeniería en función de objetivos nacionales. *1.º Congreso Venezolano de Enseñanza de la Ingeniería, Acta final*. Caracas, CIV, 1975.
——. ¿Que tipos de ingenieros se necesitan y con que prioridad? *2.º Congreso Venezolano de Enseñanza de la Ingeniería, Arquitectura, y Profesiones Afines, Valencia, Venezuela, 1977.*